FUNDAMENTALS OF DATA MINING IN GENOMICS AND PROTEOMICS

FUNDAMENTALS OF DATA MINING IN GENOMICS AND PROTEOMICS

Edited by
Werner Dubitzky
University of Ulster, Coleraine, Northern Ireland

Martin Granzow
Quantiom Bioinformatics GmbH & Co. KG, Weingarten/Baden, Germany

Daniel Berrar
University of Ulster, Coleraine, Northern Ireland

 Springer

ISBN 978-1-4419-4291-3
e-ISBN-13: 978-0-387-47509-7
e-ISBN-10: 0-387-47509-5

Printed on acid-free paper.

9 8 7 6 5 4 3 2 1

springer.com

Preface

As natural phenomena are being probed and mapped in ever-greater detail, scientists in genomics and proteomics are facing an exponentially growing volume of increasingly complex-structured data, information, and knowledge. Examples include data from microarray gene expression experiments, bead-based and microfluidic technologies, and advanced high-throughput mass spectrometry. A fundamental challenge for life scientists is to explore, analyze, and interpret this information effectively and efficiently. To address this challenge, traditional statistical methods are being complemented by methods from data mining, machine learning and artificial intelligence, visualization techniques, and emerging technologies such as Web services and grid computing.

There exists a broad consensus that sophisticated methods and tools from statistics and data mining are required to address the growing data analysis and interpretation needs in the life sciences. However, there is also a great deal of confusion about the arsenal of available techniques and how these should be used to solve concrete analysis problems. Partly this confusion is due to a lack of mutual understanding caused by the different concepts, languages, methodologies, and practices prevailing within the different disciplines.

A typical scenario from pharmaceutical research should illustrate some of the issues. A molecular biologist conducts nearly one hundred experiments examining the toxic effect of certain compounds on cultured cells using a microarray gene expression platform. The experiments include different compounds and doses and involves nearly 20 000 genes. After the experiments are completed, the biologist presents the data to the bioinformatics department and briefly explains what kind of questions the data is supposed to answer. Two days later the biologist receives the results which describe the output of a cluster analysis separating the genes into groups of activity and dose. While the groups seem to show interesting relationships, they do not directly address the questions the biologist has in mind. Also, the data sheet accompanying the results shows the original data but in a different order and somehow transformed. Discussing this with the bioinformatician again it turns out that what the biologist wanted was not clustering (*automatic* classification or *automatic* class prediction) but *supervised* classification or *supervised* class prediction.

One main reason for this confusion and lack of mutual understanding is the absence of a conceptual platform that is common to and shared by the two broad disciplines, life science and data analysis. Another reason is that data mining in the life sciences is different to that in other typical data mining applications (such as finance, retail, and marketing) because many requirements are fundamentally different. Some of the more prominent differences are highlighted below.

A common theme in many genomic and proteomic investigations is the need for a detailed understanding (descriptive, predictive, explanatory) of genome- and proteome-related entities, processes, systems, and mechanisms. A vast body of knowledge describing these entities has been accumulated on a staggering range of life phenomena. Most conventional data mining applications do not have the requirement of such a deep understanding and there is nothing that compares to the global knowledge base in the life sciences.

A great deal of the data generated in genomics and proteomics is generated in order to analyze and interpret them in the context of the questions and hypotheses to be answered and tested. In many classical data mining scenarios, the data to be analyzed are generated as a "by-product" of an underlying business process (e.g., customer relationship management, financial transactions, process control, Web access log, etc.). Hence, in the conventional scenario there is no notion of question or hypothesis at the point of data generation.

Depending on what phenomenon is being studied and the methodology and technology used to generate data, genomic and proteomic data structures and volumes vary considerably. They include temporally and spatially resolved data (e.g., from various imaging instruments), data from spectral analysis, encodings for the sequential and spatial representation of biological macromolecules and smaller chemical and biochemical compounds, graph structures, and natural language text, etc. In comparison, data structures encountered in typical data mining applications are simple.

Because of ethical constraints and the costs and time involved to run experiments, most studies in genomics and proteomics create a modest number of observation points ranging from several dozen to several hundreds. The number of observation points in classical data mining applications ranges from thousands to millions. On the other hand, modern high-throughput experiments measure several thousand variables per observation, much more than encountered in conventional data mining scenarios.

By definition, research and development in genomics and proteomics is subject to constant change – new questions are being asked, new phenomena are being probed, and new instruments are being developed. This leads to frequently changing data processing pipelines and workflows. Business processes in classical data mining areas are much more stable. Because solutions will be in use for a long time, the development of complex, comprehensive, and expensive data mining applications (such as data warehouses) is readily justified.

Genomics and proteomics are intrinsically "global" – in the sense that hundreds if not thousands of databases, knowledge bases, computer programs, and document libraries are available via the Internet and are used by researchers and developers throughout the world as part of their day-to-day work. The information accessible through these sources form an intrinsic part of the data analysis and interpretation process. No comparable infrastructure exists in conventional data mining scenarios.

This volume presents state of the art analytical methods to address key analysis tasks that data from genomics and proteomics involve. Most importantly, the book will put particular emphasis on the common caveats and pitfalls of the methods by addressing the following questions: What are the requirements for a particular method? How are the methods deployed and used? When should a method not be used? What can go wrong? How can the results be interpreted? The main objectives of the book include:

- To be acceptable and accessible to researchers and developers both in life science and computer science disciplines – it is therefore necessary to express the methodology in a language that practitioners in both disciplines understand;
- To incorporate fundamental concepts from both conventional statistics as well as the more exploratory, algorithmic and computational methods provided by data mining;
- To take into account the fact that data analysis in genomics and proteomics is carried out against the backdrop of a huge body of existing formal knowledge about life phenomena and biological systems;
- To consider recent developments in genomics and proteomics such as the need to view biological entities and processes as systems rather than collections of isolated parts;
- To address the current trend in genomics and proteomics towards increasing computerization, for example, computer-based modeling and simulation of biological systems and the data analysis issues arising from large-scale simulations;
- To demonstrate where and how the respective methods have been successfully employed and to provide guidelines on how to deploy and use them;
- To discuss the advantages and disadvantages of the presented methods, thus allowing the user to make an informed decision in identifying and choosing the appropriate method and tool;
- To demonstrate potential caveats and pitfalls of the methods so as to prevent any inappropriate use;
- To provide a section describing the formal aspects of the discussed methodologies and methods;

- To provide an exhaustive list of references the reader can follow up to obtain detailed information on the approaches presented in the book;
- To provide a list of freely and commercially available software tools.

It is hoped that this volume will (*i*) foster the understanding and use of powerful statistical and data mining methods and tools in life science as well as computer science and (*ii*) promote the standardization of data analysis and interpretation in genomics and proteomics.

The approach taken in this book is conceptual and practical in nature. This means that the presented data-analytical methodologies and methods are described in a largely non-mathematical way, emphasizing an information-processing perspective (input, output, parameters, processing, interpretation) and conceptual descriptions in terms of mechanisms, components, and properties. In doing so, the reader is not required to possess detailed knowledge of advanced theory and mathematics. Importantly, the merits and limitations of the presented methodologies and methods are discussed in the context of "real-world" data from genomics and proteomics. Alternative techniques are mentioned where appropriate. Detailed guidelines are provided to help practitioners avoid common caveats and pitfalls, e.g., with respect to specific parameter settings, sampling strategies for classification tasks, and interpretation of results. For completeness reasons, a short section outlining mathematical details accompanies a chapter if appropriate. Each chapter provides a rich reference list to more exhaustive technical and mathematical literature about the respective methods.

Our goal in developing this book is to address complex issues arising from data analysis and interpretation tasks in genomics and proteomics by providing what is simultaneously a *design blueprint, user guide*, and *research agenda* for current and future developments in the field.

As design blueprint, the book is intended for the practicing professional (researcher, developer) tasked with the analysis and interpretation of data generated by high-throughput technologies in genomics and proteomics, e.g., in pharmaceutical and biotech companies, and academic institutes.

As a user guide, the book seeks to address the requirements of scientists and researchers to gain a basic understanding of existing concepts and methods for analyzing and interpreting high-throughput genomics and proteomics data. To assist such users, the key concepts and assumptions of the various techniques, their conceptual and computational merits and limitations are explained, and guidelines for choosing the methods and tools most appropriate to the analytical tasks are given. Instead of presenting a complete and intricate mathematical treatment of the presented analysis methodologies, our aim is to provide the users with a clear understanding and practical know-how of the relevant concepts and methods so that they are able to make informed and effective choices for data preparation, parameter setting, output post-processing, and result interpretation and validation.

As a research agenda, this volume is intended for students, teachers, researchers, and research managers who want to understand the state of the art of the presented methods and the areas in which gaps in our knowledge demand further research and development. To this end, our aim is to maintain the readability and accessibility throughout the chapters, rather than compiling a mere reference manual. Therefore, considerable effort is made to ensure that the presented material is supplemented by rich literature cross-references to more foundational work.

In a quarter-length course, one lecture can be devoted to two chapters, and a project may be assigned based on one of the topics or techniques discussed in a chapter. In a semester-length course, some topics can be covered in greater depth, covering – perhaps with the aid of an in-depth statistics/data mining text – more of the formal background of the discussed methodology. Throughout the book concrete suggestions for further reading are provided.

Clearly, we cannot expect to do justice to all three goals in a single book. However, we do believe that this book has the potential to go a long way in bridging a considerable gap that currently exists between scientists in the field of genomics and proteomics on one the hand and computer scientists on the other hand. Thus, we hope, this volume will contribute to increased communication and collaboration across the disciplines and will help facilitate a consistent approach to analysis and interpretation problems in genomics and proteomics in the future.

This volume comprises 12 chapters, which follow a similar structure in terms of the main sections. The centerpiece of each chapter represents a case study that demonstrates the use – and misuse – of the presented method or approach. The first chapter provides a general introduction to the field of data mining in genomics and proteomics. The remaining chapters are intended to shed more light on specific methods or approaches.

The second chapter focuses on study design principles and discusses replication, blocking, and randomization. While these principles are presented in the context of microarray experiments, they are applicable to many types of experiments.

Chapter 3 addresses data pre-processing in cDNA and oligonucleotide microarrays. The methods discussed include background intensity correction, data normalization and transformation, how to make gene expression levels comparable across different arrays, and others.

Chapter 4 is also concerned with pre-processing. However, the focus is placed on high-throughput mass spectrometry data. Key topics include baseline correction, intensity normalization, signal denoising (e.g., via wavelets), peak extraction, and spectra alignment.

Data visualization plays an important role in exploratory data analysis. Generally, it is a good idea to look at the distribution of the data prior to analysis. Chapter 5 revolves around visualization techniques for high-dimensional data sets, and puts emphasis on multi-dimensional scaling. This technique is illustrated on mass spectrometry data.

Chapter 6 presents the state of the art of clustering techniques for discovering groups in high-dimensional data. The methods covered include hierarchical and k-means clustering, self-organizing maps, self-organizing tree algorithms, model-based clustering, and cluster validation strategies, such as functional interpretation of clustering results in the context of microarray data.

Chapter 7 addresses the important topics of feature selection, feature weighting, and dimension reduction for high-dimensional data sets in genomics and proteomics. This chapter also includes statistical tests (parametric or nonparametric) for assessing the significance of selected features, for example, based on random permutation testing.

Since data sets in genomics and proteomics are usually relatively small with respect to the number of samples, predictive models are frequently tested based on resampled data subsets. Chapter 8 reviews some common data resampling strategies, including n-fold cross-validation, leave-one-out cross-validation, and repeated hold-out method.

Chapter 9 discusses support vector machines for classification tasks, and illustrates their use in the context of mass spectrometry data.

Chapter 10 presents graphs and networks in genomics and proteomics, such as biological networks, pathways, topologies, interaction patterns, gene-gene interactome, and others.

Chapter 11 concentrates on time series analysis in genomics. A methodology for identifying important predictors of time-varying outcomes is presented. The methodology is illustrated in a study aimed at finding mutations of the human immunodeficiency virus that are important predictors of how well a patient responds to a drug regimen containing two different antiretroviral drugs.

Automated extraction of information from biological literature promises to play an increasingly important role in text-based knowledge discovery processes. This is particularly important for high-throughput approaches such as microarrays and high-throughput proteomics. Chapter 12 addresses knowledge extraction via text mining and natural language processing.

Finally, we would like to acknowledge the excellent contributions of the authors and Alice McQuillan for her help in proofreading.

Coleraine, Northern Ireland, and Weingarten, Germany   *Werner Dubitzky*
Martin Granzow
Daniel Berrar

The following list shows the symbols or abbreviations for the most commonly occurring quantities/terms in the book. In general, uppercase boldfaced letters such as **X** refer to matrices. Vectors are denoted by lowercase boldfaced letters, e.g., **x**, while scalars are denoted by lowercase italic letters, e.g., x.

List of Abbreviations and Symbols

ACE	Average (test) classification error
ANOVA	Analysis of variance
ARD	Automatic relevance determination
AUC	Area under the curve (in ROC analysis)
BACC	Balanced accuracy (average of sensitivity and specificity)
BACC	Balanced accuracy
bp	Base pair
CART	Classification and regression tree
CV	Cross-validation
Da	Daltons
DDWT	Decimated discrete wavelet transform
ESI	Electrospray ionization
EST	Expressed sequence tag
ETA	Experimental treatment assignment
FDR	False discovery rate
FLD	Fisher's linear discriminant
FN	False negative
FP	False positive
FPR	False positive rate
FWER	Family-wise error rate
GEO	Gene Expression Omnibus
GO	Gene Ontology
ICA	Independent component analysis
IE	Information extraction
IQR	Interquartile range
IR	Information retrieval
LOOCV	Leave-one-out cross-validation
MALDI	Matrix-assisted laser desorption/ionization
MDS	Multidimensional scaling
MeSH	Medical Subject Headings
MM	Mismatch
MS	Mass spectrometry
m/z	Mass-over-charge
NLP	Natural language processing
NPV	Negative predictive value
PCA	Principal component analysis
PCR	polymerase chain reaction

PCR	Polymerase chain reaction
PLS	Partial least squares
PM	Perfect match
PPV	Positive predictive value
RLE	Relative log expression
RLR	Regularized logistic regression
RMA	Robust multi-chip analysis
S2N	Signal-to-noise
SAGE	Serial analysis of gene expression
SAM	Significance analysis of gene expression
SELDI	Surface-enhance laser desorption/ionization
SOM	Self-organizing map
SOTA	Self-organizing tree algorithm
SSH	Suppression substractive hybridization
SVD	Singular value decomposition
SVM	Support vector machine
TIC	Total ion current
TN	True negative
TOF	Time-of-flight
TP	True positive
UDWT	Undecimated discrete wavelet transform
VSN	Variance stabilization normalization
$\#(\cdot)$	Counts; the number of instances satisfying the condition in $(\cdot)$
$\bar{\mathbf{x}}$	The mean of all elements in $\mathbf{x}$
χ^2	Chi-square statistic
ϵ	Observed error rate
$\epsilon_{.632}$	Estimate for the classification error in the .632 bootstrap
$\hat{y}_i$	Predicted value for y_i (i.e., predicted class label for case $\mathbf{x}_i$)
$\neg y$	Not y
Σ	Covariance
τ	True error rate
$\mathbf{x}'$	Transpose of vector $\mathbf{x}$
D	Data set
$d(x, y)$	Distance between x and y
$E(X)$	Expectation of a random variable X
$\langle k \rangle$	Average of k
L_i	i^{th} learning set
$\Re$	Set of real numbers
T_i	i^{th} test set
TR_{ij}	Training set of the ith external and j^{th} internal loop
V_{ij}	Validation set of the ith external and j^{th} internal loop
v_i	i^{th} vertex in a network

Contents

List of Contributors

Keith A. Baggerly
Department of Biostatistics and
Applied Mathematics, University of
Texas M.D. Anderson Cancer
Center, Houston, TX 77030, USA.
kabagg@wotan.mdacc.tmc.edu

Oliver Bembom
Division of Biostatistics, University
of California, Berkeley, CA 94720-
7360, USA.
bembom@berkeley.edu

Daniel Berrar
Systems Biology Research Group,
University of Ulster, Northern
Ireland, UK.
dp.berrar@ulster.ac.uk

Benjamin M. Bolstad
Department of Statistics, University
of California, Berkeley, CA 94720-
3860, USA.
bmb@bmbolstad.com

Kevin R. Coombes
Department of Biostatistics and
Applied Mathematics, University of
Texas M.D. Anderson Cancer
Center, Houston, TX 77030, USA.
krc@odin.mdacc.tmc.edu

Joaquín Dopazo
Department of Bioinformatics,
Centro de Investigación Príncipe
Felipe, E46013, Valencia, Spain.
jdopazo@cipf.es

Werner Dubitzky
Systems Biology Research Group,
University of Ulster, Northern
Ireland, UK.
w.dubitzky@ulster.ac.uk

Martin Granzow
quantiom bioinformatics GmbH &
Co. KG, Ringstrasse 61, D-76356
Weingarten, Germany.
martin.granzow@quantiom.de

Jaroslaw Harezlak
Harvard School of Public Health,
Boston, MA 02115, USA.
jharezla@hsph.harvard.edu

Milos Hauskrecht
Department of Computer Science,
and Intelligent Systems Program,
and Department of Biomedical
Informatics, University of Pittsburgh, Pittsburgh, PA 15260,
USA.
milos@cs.pitt.edu

Robert Hoffmann
Memorial Sloan-Kettering Cancer
Center, 1275 York Avenue, New
York, NY 10021, USA.
hoffmann@cbio.mskcc.org

Peter Johansson
Computational Biology and
Biological Physics Group, Depart-
ment of Theoretical Physics,
Lund University, SE-223 62, Lund,
Sweden.
peter@thep.lu.se

Kathleen F. Kerr
Department of Biostatistics,
University of Washington, Seattle,
WA 98195, USA.
katiek@u.washington.edu

Xiaochun Li
Dana Farber Cancer Institute,
Boston, Massachusetts, USA, and
Harvard School of Public Health,
Boston, MA 02115, USA.
xiaochun@jimmy.harvard.edu

James Lyons-Weiler
Department of Biomedical
Informatics, University of Pitts-
burgh, Pittsburgh, PA 15260,
USA.
lyonsweilerj@upmc.edu

Jeffrey S. Morris
Department of Biostatistics and
Applied Mathematics, University of
Texas M.D. Anderson Cancer
Center, Houston, TX 77030, USA.
jeffmo@wotan.mdacc.tmc.edu

Richard Pelikan
Intelligent Systems Program,
University of Pittsburgh, Pittsburgh,
PA 15260, USA.
pelikan@cs.pitt.edu

Maya L. Petersen
Division of Biostatistics, University
of California, Berkeley, CA 94720-
7360, USA.
mayaliv@berkeley.edu

Markus Ringnér
Computational Biology and
Biological Physics Group, Depart-
ment of Theoretical Physics,
Lund University, SE-223 62, Lund,
Sweden.
markus@thep.lu.se

Carlos Rodríguez-Caso
ICREA-Complex Systems Lab,
Universitat Pompeu Fabra (GRIB),
Dr Aiguader 80, 08003 Barcelona,
Spain.
carlos.rodriguez@upf.edu

Richard Simon
National Cancer Institute, Rockville,
MD 20852, USA.
rsimon@mail.nih.gov

Ricard V. Solé
ICREA-Complex Systems Lab,
Universitat Pompeu Fabra (GRIB),
Dr Aiguader 80, 08003 Barcelona,
Spain, and Santa Fe Institute, 1399
Hyde Park Road, NM 87501, USA.
ricard.sole@upf.edu

Michal Valko
Department of Computer Science,
University of Pittsburgh, Pittsburgh,
PA 15260, USA.
michal@cs.pitt.edu

Mark J. van der Laan
Division of Biostatistics,
University of California, Berkeley,
CA 94720-7360, USA.
laan@stat.berkeley.edu

Introduction to Genomic and Proteomic Data Analysis

Daniel Berrar[1], Martin Granzow[2], and Werner Dubitzky[1]

[1] Systems Biology Research Group, University of Ulster, Northern Ireland, UK.
dp.berrar@ulster.ac.uk, w.dubitzky@ulster.ac.uk
[2] quantiom bioinformatics GmbH & Co. KG, Ringstrasse 61, D-76356 Weingarten, Germany.
martin.granzow@quantiom.de

1.1 Introduction

Genomics can be broadly defined as the systematic study of genes, their functions, and their interactions. Analogously, *proteomics* is the study of proteins, protein complexes, their localization, their interactions, and posttranslational modifications. Some years ago, genomics and proteomics studies focused on one gene or one protein at a time. With the advent of high-throughput technologies in biology and biotechnology, this has changed dramatically. We are currently witnessing a paradigm shift from a traditionally hypothesis-driven to a data-driven research. The activity and interaction of thousands of genes and proteins can now be measured simultaneously. Technologies for genome- and proteome-wide investigations have led to new insights into mechanisms of living systems. There is a broad consensus that these technologies will revolutionize the study of complex human diseases such as Alzheimer syndrome, HIV, and particularly cancer. With its ability to describe the clinical and histopathological phenotypes of cancer at the molecular level, gene expression profiling based on microarrays holds the promise of a patient-tailored therapy. Recent advances in high-throughput mass spectrometry allow the profiling of proteomic patterns in biofluids such as blood and urine, and complement the genomic portray of diseases.

Despite the undoubted impact that these technologies have made on biomedical research, there is still a long way to go from bench to bedside. High-throughput technologies in genomics and proteomics generate myriads of intricate data, and the analysis of these data presents unprecedented analytical and computational challenges. On one hand, because of ethical, cost and time constraints involved in running experiments, most life science studies include a modest number of cases (i.e., samples), n. Typically, n ranges from several dozen to several hundred. This is in stark contrast with conventional data mining applications in finance, retail, manufacturing and engineering, for which data mining was originally developed. Here, n frequently is in the order of thousands or millions. On the other hand, modern high-throughput experiments measure several thousand variables per case, which is considerably more than in classical data mining scenarios. This problem is known as the *curse of dimensionality* or *small-n-large-p problem*. In genomic and proteomic data sets, the number of variables, p, (e.g., genes or m/z values) can be in the order of 10^4, whereas the number of cases, n, (e.g., biological specimens) is currently in the order of 10^2.

These challenges have prompted scientists from a wide range of disciplines to work together towards the development of novel methods to analyze and interpret high-throughput data in genomics and proteomics. While it is true that interdisciplinary efforts are needed to tackle the challenges, there has also been a realization that cultural and conceptual differences among the disciplines and their communities are hampering progress. These difficulties are further aggravated by continuous innovation in these areas. A key aim of this volume is to address this conceptual heterogeneity by establishing a common ontology of important notions.

Berry and Linoff (1997) define data mining broadly as *"the exploration and analysis, by automatic or semiautomatic means, of large quantities of data in order to discover meaningful patterns and rules."* In this introduction we will follow this definition and emphasize the two aspects of exploration and analysis. The exploratory approach seeks to gain a basic understanding of the different qualitative and quantitative aspects of a given data set using techniques such as data visualization, clustering, data reduction, etc. Exploratory methods are often used for hypothesis generation purposes. Analytical techniques are normally concerned with the investigation of a more precisely formulated question or the testing of a hypothesis. This approach is more confirmatory in nature. Commonly addressed analytical tasks include data classification, correlation and sensitivity analysis, hypothesis testing, etc. A key pillar of the analytical approach is traditional statistics, in particular inferential statistics. The section on basic concepts of data analysis will therefore pay particular attention to statistics in the context of small-sample genomic and proteomic data sets.

This introduction first gives a short overview of current and emerging technologies in genomics and proteomics, and then defines some basic terms and notations. To be more precise, we consider *functional genomics*, also referred to as *transcriptomics*. The chapter does not discuss the technical details of these technologies or the respective wet lab protocols; instead, we consider the basic concepts, applications, and challenges. Then, we discuss some fundamental concepts of data mining, with an emphasis on *high-throughput technologies*. Here, high-throughput refers to the ability to generate large quantities of data in a single experiment. We focus on *DNA microarrays* (transcriptomics) and *mass spectrometry* (proteomics). While this presentation is necessarily incomplete, we hope that this chapter will provide a useful framework for studying the more detailed and focused contributions in this volume. In a sense, this chapter is intended as a "road map" for the analysis of genomic and proteomic data sets and as an overview of key analytical methods for:

- data pre-processing;
- data visualization and inspection;
- class discovery;
- feature selection and evaluation;
- predictive modeling; and
- data post-processing and result interpretation.

1.2 A Short Overview of Wet Lab Techniques

A comprehensive overview of genomic and proteomic techniques is beyond the scope of this book. However, to provide a flavor of available techniques, this section briefly outlines methods that measure gene or protein *expression*.[3]

1.2.1 Transcriptomics Techniques in a Nutshell

Polymerase chain reaction (PCR) is a technique for the cyclic, logarithmic amplification of specific DNA sequences (Saiki et al., 1988). Each cycle comprises three stages: DNA denaturation by temperature, annealing with hybridization of primers to single-stranded DNA, and amplification of marked DNA sequences by polymerase (Klipp et al., 2005). Using reverse transcriptase, a cDNA copy can be obtained from RNA and used for cloning of nucleotide sequences (e.g., mRNA). This technique, however, is only semi-quantitative due to saturation effects at later PCR cycles, and due to staining with ethidium bromide. *Quantitative real-time reverse transcriptase PCR* (qRT-PCR) uses fluorescent dyes instead to mark specific DNA sequences. The increase of fluorescence over time is proportional to the generation of marked sequences (*amplicons*), so that the changes in gene expression can be monitored in real time. qRT-PCR is the most sensitive and most flexible quantification method and is particularly suitable to measure low-abundance mRNA (Bustin, 2000). qRT-PCR has a variety of applications, including viral load quantitation, drug efficacy monitoring, and pathogen detection. qRT-PCR allows the simultaneous expression profiling for approximately 1 000 genes and can distinguish even closely related genes that differ in only a few base pairs (Somogyi et al., 2002).

The *ribonuclease protection assay* (RPA) detects specific mRNAs in a mixture of RNAs (Hod, 1992). mRNA probes of interest are targeted by radioactively or biotin-labeled complementary mRNA, which hybridize to double-stranded molecules. The enzyme ribonuclease digests single-stranded mRNA,

---

[3] For an exhaustive overview of wet lab protocols for mRNA quantitation, see, for instance, Lorkowski and Cullen (2003).

so that only probes that found a hybridization partner remain. Using electrophoresis, the sample is then run through a polyacrylamide gel to quantify mRNA abundances. RPAs can simultaneously quantify absolute mRNA abundances, but are not suitable for real high-throughput analysis (Somogyi et al., 2000).

Southern blotting is a technique for the detection of a particular sequence of DNA in a complex mixture (Southern, 1975). Separation of DNA is done by electrophoresis on an agarose gel. Thereafter, the DNA is transferred onto a membrane to which a labeled probe is added in a solution. This probe bonds to the location it corresponds to and can be detected.

Northern blotting is similar to Southern blotting; however, it is a semi-quantitative method for detection of mRNA instead of DNA. Separation of mRNA is done by electrophoresis on an agarose gel. Thereafter, the mRNA is transferred onto a membrane. An oligonucleotide that is labeled with a radioactive marker is used as target for an mRNA that is run through a gel. This mRNA is located at a specific band in the gel. The amount of measured radiation in this band depends on the amount of hybridized target to the probe.

Subtractive hybridization is one of the first techniques to be developed for high-throughput expression profiling (Sargent and Dawid, 1983). cDNA molecules from the tester sample are mixed with mRNA in the driver sample, and transcripts expressed in both samples hybridize to each other. Single- and double-stranded molecules are then chromatographically separated. Single-stranded cDNAs represent genes that are expressed in the tester sample only. Moody (2001) gives an overview of various modifications of the original protocol. Diatchenko et al. (1996) developed a protocol for *suppression subtractive hybridization* (SSH), which selectively amplifies differentially expressed transcripts and suppresses the amplification of abundant transcripts. SSH includes PCR, so that even small amounts of RNA can be analyzed. SSH, however, is only a qualitative technique for comparing relative expression levels in two samples (Moody, 2001).

In contrast to SSH, the *differential display technique* can detect differential transcript abundance in more than two samples, but is also unable to measure expression quantitatively (Liang and Pardee, 1992). First, mRNA is reverse-transcribed to cDNA and amplified by PCR. The PCR clones are then labeled, either radioactively or using a fluorescent marker, and electrophoresed through a polyacrylamide gel. The bands with different intensities represent the transcripts that are differentially expressed in the samples.

Serial analysis of gene expression (SAGE) is a quantitative and high-throughput technique for rapid gene expression profiling (Velculescu et al., 1995). SAGE generates double-stranded cDNA from mRNA and extracts short sequences of 10-15 bp (so-called *tags*) from the cDNA. Multiple sequence tags are then concatenated to a double-stranded stretch of DNA, which is then amplified and sequenced. The expression profile is determined based on the abundance of individual tags.

A major breakthrough in high-throughput gene expression profiling was reached with the development of *microarrays* (Schena et al., 1995). Arguably, spotted cDNA arrays, spotted and *in situ* synthesized chips currently represent the most commonly used array platforms for assessing mRNA transcript levels. cDNA chips consist of a solid surface (nylon or glass) onto which probes of nucleotide sequences are spotted in a grid-like arrangement (Murphy, 2002). Each spot represents either a gene sequence or an *expressed sequence tag* (EST). cDNA microarrays can be used to compare the relative mRNA abundance in two different samples. In contrast, in situ synthesized oligonucleotide chips such as Affymetrix GeneChips measure absolute transcript abundance in one single sample (more details can be found in Chapter 3).

1.2.2 Proteomics Techniques in a Nutshell

In *Western blotting*, protein-antibody complexes are formed on a membrane, which is incubated with an antibody of the primary antibody. This secondary antibody is linked to an enzyme triggering a chemiluminescence reaction (Burnette, 1981). Western blotting produces bands of protein-antibody-antibody complexes and can quantify protein abundance absolutely.

Two-dimensional polyacrylamide gel electrophoresis (2D-PAGE) separates proteins in the first dimension according to charge and in the second dimension according to molecular mass (O'Farrell, 1975). 2D-PAGE is a quantitative high-throughput technique, allowing a high-resolution separation of over 10 000 proteins (Klose and Kobalz, 1995). A problem with 2D-PAGE is that high-abundance proteins can co-migrate and obscure low-abundance proteins (Honoré et al., 2004). *Two-dimensional difference in-gel electrophoresis* (2D-DIGE) is one of the many variations of this technique (Unlu et al., 1997). Here, proteins from two samples (e.g., normal vs. diseased) are differentially labeled using fluorescent dyes and simultaneously electrophoresed.

Mass spectrometry (MS) plays a pivotal role in the identification of proteins and their post-translational modifications (Glish and Vachet, 2003; Honoré et al., 2004). Mass spectrometers consist of three key components: (*i*) An *ion source*, converting proteins into gaseous ions; (*ii*) a *mass analyzer*, measuring the *mass-to-charge ratio* (m/z) of the ions, and (*iii*) a *detector*, counting the number of ions for each m/z value. Arguably the two most common types of ion sources are *electrospray ionization* (ESI) (Yamashita and Fenn, 1984) and *matrix-assisted laser desorption/ionization* (MALDI) (Karas et al., 1987). Glish and Vachet (2003) give an excellent overview of various mass analyzers that can be coupled with these ion sources. The *time-of-flight* (TOF) instrument is arguably the most commonly used analyzer for MALDI. In short, the protein sample is mixed with matrix molecules and then crystallized to spots on a metal plate. Pulsed laser shots to the spots irradiate the mixture and trigger ionization. Ionized proteins fly through the ion chamber and hit the detector. Based on the applied voltage and ion velocity, the m/z of each ion can be determined and displayed in a spectrum. *Surface-enhanced laser desorption/ionization time-of-flight* (SELDI-TOF) is a relatively new variant of MALDI-TOF (Issaq et al., 2002; Tang et al., 2004). A key element in SELDI-TOF MS is the protein chip with a chemically treated surface to capture classes of proteins under specific binding conditions. MALDI- and SELDI-TOF MS are very sensitive technologies and inherently suitable for high-throughput proteomic profiling. The pulsed laser shots usually generate singularly protonated ions $[M+H]^+$; hence, a sample that contains an abundance of a specific protein should produce a spectrum where the m/z value corresponding to this protein has high intensity, i.e., stands out as a peak. However, mass spectrometry is inherently semi-quantitative, since protein abundance is not measured directly, but via ion counts. Chapter 4 provides more details about these technologies.

2D-PAGE and mass spectrometry are currently the two key technologies in proteomic research. Further techniques include: (*i*) *Yeast two-hybrid*, an *in vivo* technique for deciphering protein-protein interactions (Fields and Song, 1989); (*ii*) *phage display*, a technique to determine peptide- or domain-protein interactions (Hoogenboom et al., 1998); and (*iii*) *peptide* and *protein chips*, comprising affinity probes, i.e., reagents such as antibodies, antigens, recombinant proteins, arrayed in high density on a solid surface (MacBeath, 2002). Similarly to two-color microarray experiments, the probes on the chip interact with their fluorescently labeled target proteins, so that captured proteins can be detected and quantified. Three major problems currently hamper the application of protein chips: The production of specific probes, the affixation of functionally intact proteins on high-density arrays, and cross-reactions of antibody reagents with other cellular proteins.

1.3 A Few Words on Terminology

Arguably the most important interface between wet lab experiments in genomics and proteomics and data mining is data. We could summarize this via a logical workflow as follows: *Wet lab experiments → data → data mining.* In this section, we briefly outline some important terminology often used in genomics and proteomics to capture, structure, and characterize data.

A *model* refers to the instantiation of a mathematical representation or formalism, and reflects a simplified entity in the real world. For example, a particular decision tree classifier that has been constructed using a specific decision tree learning algorithm based on a particular data set is a model. Hence, identical learning algorithms can lead to different models, provided that different data subsets are used.

The terms *probe* and *target* sometimes give rise to confusion. In general, probe refers to the substance that interacts in a selective and predetermined way with the target substance so as to elicit or measure a specific property or quantity. In genomics and proteomics, the term "probe" is nowadays used for substances or molecules (e.g., nucleic acids) affixed to an array or chip, and the term "target" designates the substances derived from the studied samples that interact with the probe.

The terms *feature, variable,* and *attribute* are also widely used as synonyms. The term *target variable* (or simply *target*) is often used in machine learning and related areas to designate the class label in a classification scenario. The statistical literature commonly refers to this as the *response* or *dependent variable,* whereas the features are the *predictors, independent variables,* or *covariates.*

In genomics and proteomics the terms *profile, signature, fingerprint,* and others are often used for biologically important data aggregates. Below, we briefly illustrate some of these aggregates.

In DNA microarray data analysis, the biological entity of interest is mRNA abundance. These abundances are either represented as ratio values (in cDNA chips) or absolute abundances (in oligonucleotide chips). In mass spectrometry, the biological entity of interest is the abundance of peptides/proteins or protein fragments. Provided that the pulsed laser shots generate singularly protonated ions, a specific peptide/protein or protein fragment is represented by a specific m/z value. The ions corresponding to a specific m/z value are counted and used as a measure of protein abundance.

A *gene expression profile* is a vector representing gene expression values relating to a single gene across multiple cases or conditions. The term *gene expression signature* is commonly used synonymously. A (gene) *array profile* is a vector that describes the gene expression values for multiple genes for a single case or under a single condition.

For mass spectrometry data, a *"protein expression profile"* is a vector representing the intensity (i.e., ion counts) of a single m/z value across multiple cases or conditions.[4] A *mass spectrum* is a vector that describes the intensity of multiple m/z values for a single case or under a single condition. Figure 1.1 shows the conceptually similar microarray matrix and MS matrix.

1.4 Study Design

High-throughput experiments are often of an exploratory nature, and highly focused hypotheses may not always be desired or possible. Of critical importance, however, is that the objectives of the analysis are precisely specified before the data are generated. Clear objectives guide the study design, and flaws at this stage cannot be corrected by data mining techniques. *"Pattern recognition and data mining are often what you do when you don't know what your objectives are."* (Simon, 2002). Chapter 2 of this volume addresses the experimental study design issues. The design principles are discussed in the

---

[4] Note that in general, a specific m/z value cannot be directly mapped to a specific protein, because the mass is not sufficient to identify a protein. See the discussion on peak detection and peak identification in Chapter 4, pages 81–82.

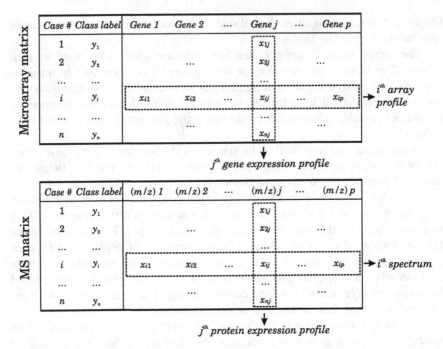

Fig. 1.1. Microarray matrix and mass spectrometry matrix. In the microarray matrix, x_{ij} is the expression value of the jth gene of the ith array. In the MS matrix, x_{ij} refers to the intensity of the jth m/z value of the ith spectrum.

context of microarray experiments, but also apply to other types of experiments.

Normally, a study is concerned with one or more scientific questions in mind. To answer these questions, a rational study design should identify which analytical tasks need to be performed and which analytical methods and tools should be used to implement these tasks. This mapping of *question → task → method* is the first hurdle that needs to be overcome in the data mining process.

1.5 Data Mining

While there is an enormous diversity of data mining methodologies, methods and tools, there are a considerable number of principle concepts, issues and techniques that appear in one form or another in many data mining applications. This section and its subsections try to cover some of these notions. Figure 1.2 depicts a typical "analysis pipeline", comprising five essential phases after the study design. The following sections describe this pipeline in more details, placing emphasis on class comparison and class discrimination. Chapter 6 discusses class discovery in detail.

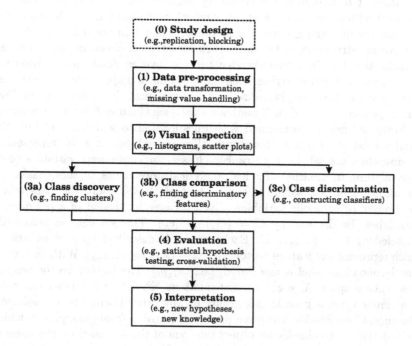

Fig. 1.2. A typical "data mining pipeline" in genomics and proteomics.

1.5.1 Mapping Scientific Questions to Analytical Tasks

Frequently asked questions in genomic and proteomic studies include:

1. Are there any interesting patterns in the data set?
2. Are the array profiles characteristic for the phenotypes?
3. Which features (e.g., genes) are most important?

To formulate the first question more precisely is already a challenge. What is meant by a "pattern", and how should one measure "interestingness"? A pattern can refer to groups in the data. This question can be translated into a clustering task. Informally, clustering is concerned with identifying meaningful groups in the data, i.e., a convenient organization and description of the data. Clustering is an unsupervised learning method as the process is not guided by pre-defined class labels but by similarity and dissimilarity of cases according to some measure of similarity. Clustering refers to an exploratory approach to reveal relationships that may exist in the data, for instance, hierarchical topologies. There exists a huge arsenal of different clustering methods. They have in common that they all ultimately rely on the definition of a measure of similarity (or, equivalently, dissimilarity) between objects. Clustering methods attempt to maximize the similarity between these objects (i.e., cases or features) within the same group (or cluster), while minimizing the similarity between the different groups. For instance, if one is interested in identifying hierarchical structures in the data, then hierarchical clustering methods can organize the data into tree-like structures known as *dendrogram*. Adopting this approach, the underlying scientific question is mapped into a clustering task, which, in this case, is realized via a hierarchical clustering method. Various implementations of such methods exist (see Chapter 6 for an overview).

Many studies are concerned with questions as to whether and how the profiles relate to certain phenotypes. The phenotypes may be represented by discrete class labels (e.g., cancer classes) or continuous variables (e.g., survival time in months). Typical analytical approaches to these tasks are *classification* or *regression*. In the context of classification, the class labels are discrete or symbolic variables. Given n cases, let the set of k pre-defined class labels be denoted by $C = \{c_1, c_2, \ldots c_k\}$. This set can be arbitrarily relabeled as $Y = \{1, 2, \ldots k\}$. Each case $\mathbf{x}_i$ is described by p observations, which represent the feature vector, i.e., $\mathbf{x}_i = (x_{i1}, x_{i2}, \ldots x_{ip})$. With each case, exactly one class label is associated, i.e., $(\mathbf{x}_i, y_i)$. The feature vector belongs to a feature space X, e.g., the real numbers $\Re^p$. The class label can refer to a tumor type, a genetic risk group, or any other phenotype of biological relevance. Classification involves a process of learning-from-examples, in which the objective is to classify an object into one of the k classes on the basis of an observed measurement, i.e., to predict y_i from $\mathbf{x}_i$.

The task of regression is closely related to the task of classification, but differs with respect to the class variables. In regression, these variables are continuous values, but the learning task is similar to the aforementioned mapping function. Such a continuous variable of interest can be the survival outcome of cancer patients, for example. Here, the regression task may consist in finding the mapping from the feature vector to the survival outcome.

A plethora of sophisticated classification/regression methods have been developed to address these tasks. Each of these methods is characterized by a set of idiosyncratic requirements in terms of data pre-processing, parameter configuration, and result evaluation and interpretation.

It should be noted that the second question mentioned in the beginning of Section 1.5.1 does not translate into a clustering task, and hence clustering methods are inappropriate. Simon (2005) pointed out that one of the most common errors in the analysis of microarray data is the use of clustering methods for classification tasks.

The *No Free Lunch theorem* suggests that no classifier is inherently superior to any other (Wolpert and Macready, 1997). It is the type of the problem and the concrete data set at hand that determines which classifier is most appropriate. In general, however, it is advisable to prefer the simplest model that fits the data well. This postulate is also known as *Occam's razor*. Somorjai et al. (2003) criticized the common practice in classifying microarray data that does not respect Occam's razor. Frequently, the most sophisticated models are applied. Currently, *support vector machines* (SVMs) are considered by many as one of the most sophisticated techniques. Empirical evidence has shown that SVMs perform remarkably well for high-dimensional data sets involving two classes, but in theory, there are no compelling reasons why SVMs should have an edge on the curse of dimensionality (Hastie et al., 2002). Comparative studies have demonstrated that simple methods such as nearest-neighbor classifiers often perform as well as more sophisticated methods (Dudoit et al., 2002).

More important than the choice of the classifier is its correct application. To assess whether the array profiles, for instance, are characteristic for the phenotypes, it is essential to embed the construction and application of the classifier in a solid statistical framework. Section 1.5.5 and Chapter 8 discuss this issue in detail.

With respect to class discrimination, we are interested in those features that differ significantly among the different classes. Various methods for feature weighting and selection exist. Chapter 7 presents the state of the art of feature selection techniques in the context of genomics and proteomics. Feature selection is closely linked to the construction of a classifier, because in general, classification performance improves when non-discriminatory features are discarded.

It is important to be clear about the analysis tasks, because they may dictate what to do next in the data pre-processing step. For instance, if the task is tackled by a hierarchical clustering method, then missing values in the data set need to be handled. Some software packages may not be able to perform clustering if the data set has missing values. In contrast, if the problem is identified as a classification task and addressed by a model that is inherently able to cope with missing values (e.g., some types of decision trees), then missing values do not necessarily need to be replaced.

1.5.2 Visual Inspection

There are many sources causing artifacts in genomics and proteomics data sets that may be confused as real measurements. High-throughput genomic and proteomic data sets are the result of a complex scientific instrument, comprising laboratory protocols, technical equipment and the human element. The human eye is an invaluable tool that can help in quality assessment of data. Looking at the data distribution prior to analysis is often a highly valuable exercise.

Many parametric methods (e.g., the standard *t*-test) assume that the data follows approximately a normal distribution. Histogram plots like those shown in Figure 1.3a can reveal whether the normality assumption is violated. Alternatively, a statistical test for normality (e.g., Anderson-Darling test) may be used. To coerce data into a normality distribution the data may need to be transformed prior to applying a method requiring normality. Figure 1.3a shows the frequency distribution of a two-color microarray experiment based on cDNA chips, which represents expression values as intensity ratios.

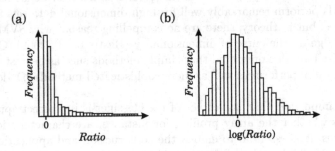

Fig. 1.3. Frequency distribution of (a) intensity ratios, and (b) log-transformed intensity ratios.

In Figure 1.3a, a large proportion of the values are confined to the lower end of the observed scale. This is referred to as positive skewness of the ratio data. Here, values indicating underexpression are "squashed" in the interval $(0, 1)$. A simple log-transformation usually provides for a good approximation of the normal distribution (see Figure 1.3b).

Data integration has become a buzzword in genomics and proteomics. However, current research practice is characterized by multiple array platforms and protocols, and even expression data from the same tissue type are not directly comparable when they originate from different platforms (Morris et al., 2003). This problem is exacerbated when data are pooled across different laboratories. Prior to integrating data, it may be useful to inspect the data using *multidimensional scaling* (MDS) or *principal component analysis* (PCA) (see Chapter 5).

Figure 1.4 shows a score plot of the first and second principal components of two contrived microarray experiments generated by two different laboratories (marked by □ and •, respectively). In this example, the largest source of variation (reflected by the first principal component) is due to (unknown) laboratory peculiarities; hence, the expression values in the two data sets are not directly comparable.

Visual inspection is not only useful prior to data analysis, but should accompany the entire analysis process. The visual examination of data analysis steps by meaningful visualization techniques supports the discovering of mistakes, e.g., when a visualization does not appear the way we expected it to look like. Furthermore, visualizing the single analysis steps fosters the confidence in the data mining results.

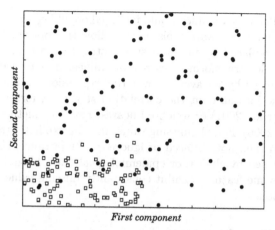

Fig. 1.4. Score plot of the first and second principal component.

1.5.3 Data Pre-Processing

Pre-processing encompasses a wide range of methods and approaches that make the data amenable to analysis. In the context of microarrays, data pre-processing includes the acquisition and processing of images, handling of missing values, data transformation, and filtering. Chapter 3 addresses these issues in detail. In data sets based on MALDI/SELDI-TOF MS, pre-processing includes identification of valid m/z regions, spectra alignment, signal denoising or smoothing, baseline correction, peak extraction, and intensity normalization (see Chapter 4 for details). Precisely which pre-processing needs to be done depends on the analytical task at hand.

1.5.3.1 Handling of Missing Values

Genomic and proteomic data sets can exhibit missing values for various reasons. For instance, missing values in microarray matrices can be due to problems in image resolution, dust and scratches on the array, and systematic artifacts from robotic printing. Essentially, there exist four different approaches for coping with missing values.

First, if the number of missing values is relatively small, then we might discard entire profiles. This would be the most obvious, albeit drastic, solution. Second, missing values are ignored because the data mining methods to be used are intrinsically able to cope with them. Some decision tree algorithms, for instance, are able to cope with missing values automatically. Methods that compute pair-wise distances between objects (e.g., clustering algorithms) could discard pairs where one partner is missing. For instance, suppose that the value x_{ij} depicted in the data matrix in Figure 1.1 is missing, and the distance between the jth and the $(j+1)^{th}$ expression profile is to be computed. Then the distance would be based on all (x_{kj}, x_{kj+1}) with $k \neq i$. Unfortunately, many software tools do not allow this option. Third, missing values may be replaced by imputed substitutes. In the context of microarray matrices of log-transformed expression values, missing values are often replaced by zero or by an average over the expression profile. More robust approaches take into account the correlation structure, for example, simple (Troyanskaya et al., 2001) or weighted nearest-neighbor methods (Johansson and Hakkinen, 2006). Fourth, missing values may be explicitly treated as missing information (i.e., not replaced or ignored). For instance, consider a data set that is enriched by clinical or epidemiological data. Here, it might be interesting that some features exhibit consistently missing values in subgroups of the population.

1.5.3.2 Data Transformations

Data transformation includes a wide range of techniques. Transformation to normality refers to the adjustment of the data so that they follow approximately a normal distribution.[5] Figure 1.3 showed an example of log-transformation.

Ideally, a numerical value in the expression matrix reflects the true level of transcript abundance (e.g., in oligonucleotide chips), some abundance ratio (e.g., in cDNA chips), or protein abundance (e.g., in mass spectrometry). However, due to imperfections of instruments, lab conditions, materials, etc. the measurements deviate from the true expression level. Such deviations are referred to as measurement errors and can be decomposed into two elements, *bias* and *variance*.

The measurement error due to variance (random error) is often normally distributed, meaning that deviations from the true value in either direction are equally frequent, and that small deviations are more frequent than large ones. A standard way of addressing this class of error is experiment replication. A well-designed study is of paramount importance here. Chapter 2 deals with this topic in more detail.

The bias describes the systematic error of the instrument and measurement environment. The goal of data normalization is to correct for the systematic errors and adjust the data for subsequent analysis. There exist various sources of systematic errors, for instance:

- *Experimenter bias*: Experiments carried out by the same person can cluster together. In microarray data, this has been identified as one of the largest sources of bias (Morrison and Hoyle, 2002).

---

[5] Log-transformation, albeit commonly applied in microarray data analysis, is not free from problems. James-Lyons Weiler, for example, argues that this transformation can entail a considerable loss of information in case-control studies (http://bioinformatics.upmc.edu/Help/Recommendations.html).

- *Variability in experimental conditions*: Factors such as temperature, date, and sequence can have an effect on the experiment.
- *Sample collection and preparation*: Probe processing can affect the experiment.
- *Machine parameters*: Machine calibration (e.g., scanner settings) can change over time and impact the experiment.

Data re-scaling refers to the experiment-wise transformation of the data in such a way that their variances become comparable. For example, the values resulting from two hybridizations can have different variances, making their comparison more difficult. Particularly, when the values are averaged over multiple hybridization replicates, the variances of the individual hybridizations should be equal, so that each replicate contributes an equal amount of information to the average. The z-score transformation rescales a variable by subtracting the mean from each value and then dividing by its standard deviation. The resulting z-scores are normally distributed with mean 0 and standard deviation 1. This z-score transformation can also be applied for per-feature scaling, so that the mean of each feature over multiple cases equals 0 and the standard deviation equals 1. The gene-wise re-scaling may be appropriate prior to some analytical tasks, e.g., clustering. Hedenfalk et al. (2003), for example, pre-processed the expression values by computing the z-scores over the samples.

Which data transformation method should be performed on a concrete data set at hand? This question does not have a definite answer. For microarray data, intricate normalization techniques exist, for example, methods that rely on regression techniques (Morrison and Hoyle, 2002). In general, it is good to keep the raw data and to maintain an audit trail of the performed data transformations, with the specific parameter settings. Chapter 3 discusses normalization issues in the context of microarrays, Chapter 4 in the context of MALDI/SELDI-TOF MS data.

1.5.4 The Problem of Dimensionality

The small-n-large-p problem represents a major challenge in high-throughput genomic and proteomic data sets. This problem can be addressed in two different ways: (i) By projecting the data onto a lower-dimensional space, i.e., by replacing the original data by surrogate features, and (ii) by selecting a subset of the original features only.

1.5.4.1 Mapping to Lower Dimensions

Principal component analysis (PCA, a.k.a. *Karhunen-Loève transform*) based on *singular value decomposition* (SVD) is an unsupervised technique to detect and replace linear redundancies in data sets. PCA defines a set of hybrid or surrogate features (*principal components*) that are composites of the original features. These new features are guaranteed to be linearly independent and non-redundant. It is noteworthy, however, that non-linear dependencies may still exist. PCA accounts for as much of the variation in the original data by as few as possible new features (see Chapter 5).

An important caveat should be taken into consideration. Suppose that the data set comprises only two expression profiles. Assume that the variance of one profile is much larger than the variance of the other one, but both are equally important for discriminating the classes. In this scenario, the first principal component will be dominated by the expression profile of the first gene, whereas the profile of the second feature has little influence. If this effect is not desired, then the original values should be re-scaled to mean 0 and variance 1 (z-score transformation). For example, it is generally advisable to standardize the expression values of time series data, because we are generally more interested in how the expression of a gene varies over time than in its steady-state expression level. PCA can also be based on the correlation matrix instead of the covariance matrix. This approach accounts for an unequal scaling of the original variables. Computing the principal components based on the correlation matrix is equivalent to computing the components based on the covariance of the standardized variables.

In numerous studies PCA has proven to be a useful dimension reduction technique for microarray data analysis, for instance, Alter et al. (2000); Raychaudhuri et al. (2000). *Independent component analysis* (ICA) is a technique that extracts statistically independent patterns from the data and, in contrast to PCA, does not search for uncorrelated features.

It should be noted that PCA is an unsupervised method, i.e., it does not make use of the class labels. Alternatively, *partial least squares* (PLS) regression is a supervised method that produces surrogate features (*latent vectors*) that explain as much as possible of the covariance between the class labels and the data (Hastie et al., 2002).

The biological interpretation of the hybrid features produced by PCA is not trivial. For example, the first eigengene captures the most important global pattern in the microarray matrix, but the numerical values cannot be interpreted as (ratios of) mRNA abundances any more. In contrast, the interpretation of weighted original features is obvious.

1.5.4.2 Feature Selection and Significance Analysis

Feature selection aims at selecting the relevant features and eliminating the irrelevant ones. This selection can be achieved either explicitly by selecting a subset of "good" features, or implicitly by assigning weights to all features, where the value of the weight corresponds to the relative importance of the respective feature. Implicit feature selection is also called *feature weighting*. The following four issues are relevant for all explicit feature selection procedures:

1. How to begin the search?

   Basically, there exist two main strategies: In *forward selection*, the heuristic starts with an empty set and iteratively adds relevant features. In *backward elimination*, the heuristic starts with all features and iteratively eliminates the irrelevant ones (e.g., *Markov blanket filtering*).

2. How to explore the data space?

   Here, the question is which feature should be evaluated next. In the simplest way, the features are evaluated sequentially, i.e., without preference in terms of order.

3. How to evaluate a feature?

   Here, the issue is how the discriminating power is to be measured.

4. When to stop the search?

   The number of relevant features can be determined by a simple thresholding, e.g., by limiting the number of discriminating features to, say, 20 per class, or by focusing on all features that are significantly different.

1.5.4.3 Test Statistics for Discriminatory Features

There exist various metrics for feature weighting; Chapter 7 gives an overview. The two-sample t-statistic (for unpaired data) is one of the most commonly used measures to assess the discriminatory power of a feature in a two-class scenario. Essentially, this statistic is used to test the hypothesis whether two sample means are equal. The two-sample t-statistic for unequal[6] variances is given in Equation 1.1.

$$T = \frac{m_1 - m_2}{\sqrt{\frac{s_1^2}{n_1} + \frac{s_2^2}{n_2}}} \,, \tag{1.1}$$

where m_1 is the mean expression value of the feature in class #1, m_2 is the mean expression in class #2, n_1 and n_2 are the number of cases in class #1 and #2, respectively; s_1^2 and s_2^2 are the variances in class #1 and #2, respectively, and the degrees of freedom are estimated using the approximation by *Welch-Satterthwaite*.[7]

Assuming that the feature values follow approximately a normal distribution, the t-statistic can be used for testing the null hypothesis that the mean expression value of the feature is equal in the two classes. Note that the null hypothesis, H_0, of equal mean expression, i.e., $H_0 : \mu_1 = \mu_2$, involves a two-sided test.[8] The alternative hypothesis is that either $\mu_1 > \mu_2$ or $\mu_1 < \mu_2$. The null hypothesis can be rejected if the statistic exceeds a critical value,

---

[6] Note that in general, equal variances should not be assumed. To test whether the variances are equal, Bartlett's test can be applied if the data follow a normal distribution (Bartlett, 1937); Levene's test is an alternative for smaller sample sizes and does not rely on the normality assumption (Levene, 1960).

[7] $df = (\frac{s_1^2}{n_1} + \frac{s_2^2}{n_2})^2 / (\frac{s_1^4}{n_1^2(n_1-1)} + \frac{s_2^4}{n_2^2(n_2-1)})$

[8] The population mean, μ, and variance, σ^2, are estimated by the sample mean, m, and variance, s^2, respectively.

i.e., if $|T| > t_{\frac{1}{2}\alpha, df}$. For instance, the critical value for the two-sided test at $\alpha = 0.05$ and $df = 9$ is $t \approx 2.26$. Hence, if $T > 2.26$, then we can say with 95% confidence that in class #1, the values of the feature are significantly higher than in class #2 (and vice versa, if $T < -2.26$).

A quite popular variant of the t-statistic is the *signal-to-noise* (S2N) *ratio*, introduced by Golub et al. (1999) in the context of microarray data. This metric is also known as a *Fisher-like score* (see Chapter 7, Equation (7.1), page 151, for Fisher score)[9] and expresses the discriminatory power of a feature by the difference of the empirical means m_1 and m_2, divided by the sum of their variances. This scoring metric can be easily extended to more than two classes using a one-versus-all approach. For instance, in order to compute the discriminatory power of a feature with respect to class #1, the empirical mean of this class is compared to the average of all cases that do *not* belong to class #1. However, we note that this approach is not adequate for assessing whether the sample means are significantly different. For example, assume that a data set contains five classes with ten cases each, and only one feature. Is the feature significantly different between the classes? It might be tempting to use a two-sample t-test for each possible comparison. For n classes, this would result in a total of $\frac{1}{2}n(n-1)$ pair-wise comparisons. If we specify $\alpha = 0.05$ for each individual test, then the probability of avoiding the Type I error is 95%.[10] Assume that the individual tests are independent. Then the probability of avoiding the Type I error on *all* tests is $(1-\alpha)^n$, and the probability of committing the Type I error is $1 - (1-\alpha)^n$, which is 0.40 in this example.[11]

The appropriate statistical approach to the problem in this example is the *one-way analysis of variance* (ANOVA), which tests whether the means of multiple samples are significantly different. The basic idea of this test is that under the null hypothesis (i.e., there exist no difference of means), the variance based on within-group variability should be equal to the variance based on the between-groups variability. The F-test assesses whether the ratio of these two variance estimates is significantly greater than 1. A significant result, however, only indicates that at least two sample means are different. It does not tell us which specific pair(s) of means are different. Here, it is necessary

---

[9] Note that Golub et al. (1999) use a variant of the "true" Fisher score. The difference is that the numerator in the "true" Fisher score is squared, whereas in the Fisher-like score, it is not.

[10] A Type I error (false positive) exists when a test incorrectly indicates that it has found a positive (i.e., significant) result where none actually exists. In other words, a Type I error can be thought of as an incorrect rejection of the null hypothesis, accepting the alternative hypothesis even though the null hypothesis is true.

[11] In fact, this probability is even larger, because the independence assumption is violated: If we know the difference between m_1 and m_2 and between m_1 and m_3, then we can infer the difference between m_1 and m_3; hence, only two of three differences are independent. Consequently, only two of three pair-wise comparisons are independent.

to apply post-hoc tests (such as *Tukey's, Dunnett's,* or *Duncan's test*), which take into account that more than two classes were compared with each other. The ANOVA F-test can be extended to more than one feature. However, it is necessary that the number of features (p) is greater than the number of cases (n); a "luxury" hardly met in real-world genomics and proteomics data sets. Furthermore, note that the ANOVA F-test assumes that the variances of a feature in the different classes are equal. If this is not the case, then the results can be seriously biased, particularly when the classes have a different number of cases (Chen et al., 2005).[12] However, if the classes do have equal variances, then the ANOVA F-test is the statistic of choice for comparing class means (Chen et al., 2005). There exist various alternatives to the ANOVA F-test, including *Brown and Forsythe* (Brown and Forsythe, 1974), *Welch* (Welch, 1951), *Cochran* (Cochran, 1937), and *Kruskal-Wallis* test statistic (Kruskal and Wallis, 1952). Chen et al. (2005) compared these statistics with the ANOVA F-test in the context of multiclass microarray data and observed that Brown-Forsythe, Welch, and Cochran statistics are to be preferred over the F-statistic for classes of unequal sizes and variances.

It is straightforward to convert these statistics into p-values, which have a more intuitive interpretation. The p-value is the probability of the test statistic being at least as extreme as the one observed, given that the null hypothesis is true (i.e., that the mean expression is equal between the classes). Figure 1.5 illustrates the relationship between the test statistic and the p-value for Student's t-distribution.

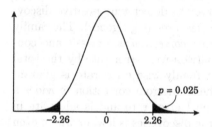

Fig. 1.5. Probability density function for Student's t-distribution and critical values for T for nine degrees of freedom.

For each class, the features can be ranked according to their p-values in ascending order and the top $x\%$ could be selected for further analysis.

1.5.4.4 Multiple Hypotheses Testing

The Type I error rate can be interpreted as the probability of rejecting a truly null hypothesis, whereas the Type II error rate is the probability of not rejecting a false null hypothesis. Feature selection based on feature weighting can be regarded as *multiple hypotheses testing*. For each feature, the null

---

[12] The ANOVA F-test applied in a two-class scenario is equivalent to the two-sample t-test assuming equal variances.

hypothesis is that there exists no significant difference between the classes (for instance, a gene is not differentially expressed). The alternative hypothesis is that it is significantly different. Adopting the terminology by Storey and Tibshirani (2003), a feature is called *truly null* if the null hypothesis is in fact true, and a feature is called *truly alternative* if the alternative hypothesis is true.

It is essential to distinguish between two different error rates: The *false positive rate* (FPR) is the rate that truly null hypotheses are rejected. The *false discovery rate* (FDR) is the rate that rejected null hypotheses are truly null. For instance, a FPR of $n\%$ implies that on average, $n\%$ of the truly null hypotheses are rejected. In contrast, a FDR of $n\%$ implies that among all rejected hypotheses, $n\%$ can be expected to be truly null.

Genomic and proteomic data sets involve testing multiple hypotheses, for instance, one significance test per feature for selecting marker genes. The Type I error rate of each individual test is the *comparison-wise error rate*, while the *family-wise error rate* (a.k.a. overall Type I error rate), is made up of the individual comparisons. Choosing a traditional p-value cut-off of $\alpha = 0.01$ or $\alpha = 0.05$ for each feature would result in an abundance of false positive discoveries. For instance, suppose that a data set contains 10 000 features. A comparison-wise error rate of 0.01 implies $0.01 \times 10\,000 = 100$ Type I errors, which means that we can expect 100 false positive discoveries. To avoid non-reproducible positive results, it is therefore necessary to adjust for multiple testing.

Reducing the Type I error rate comes at the price of an increased Type II error rate, which implies a reduced power to detect true positive discoveries. Suppose that a data set contains m features (e.g., genes). The family then includes m hypotheses. The *Bonferroni correction* is a classic and conservative approach that divides the comparison-wise error rate by the total number of comparisons, α/m; hence, the family-wise error rate is guaranteed to be smaller than or equal to α. The Bonferroni correction provides a stringent criterion for controlling the Type I error rate and is adequate in scenarios where the expected number of true discoveries is low, or where even a small number of false discoveries cannot be tolerated. On the other hand, its conservativeness entails a high Type II error rate if many discoveries are expected.

Various alternatives have been suggested to provide a better trade-off between the number of true and false positive discoveries. Manly et al. (2004) provide an excellent review of various tests for multiple comparisons, which differ with respect to their degree of conservativeness. The *Holm test* rejects more false null hypotheses than the Bonferroni method, hence has greater power (Holm, 1979). The p-values are first ranked in ascending order. The smallest p-value is then compared to α/m. If the null hypothesis cannot be rejected, then no further comparisons are made. Otherwise, the test proceeds with checking the second smallest p-value against $\alpha/(m-1)$. If the null hypothesis cannot be rejected, then the test stops; otherwise, the third smallest

p-value is compared to $\alpha/(m-2)$, and so on. *Hochberg's test* (Hochberg, 1988) rejects the null hypotheses associated with the k smallest p-values if $p_k \leq \alpha/(m-k+1)$. This test is slightly more powerful than Holm's test. A yet less conservative test has been developed by Benjamini and Hochberg (1995). This test rejects the null hypotheses associated with the smallest k p-values if $m \cdot p_k/k \leq \alpha$. Assuming that all hypotheses are truly null, $m \cdot p_k$ is the expected number of false positive discoveries. However, this number is generally overestimated for real data sets where some null hypotheses are in fact false (for instance, some genes are truly differentially expressed in the classes).

Ideally, the test by Benjamini and Hochberg would use the number of true null hypotheses, m_0, instead of m, the total number of hypotheses. However, m_0 is unknown for real-world data sets. Using the empirical distribution of the p-values, the procedure by Storey and Tibshirani (2003) replaces m by an estimate of the number of truly null hypotheses and provides for a sensible trade-off between FPR and FDR.

Which adjustment is the method of choice? Manly et al. (2003) conclude that more liberal approaches are to be preferred in exploratory genomics studies where many features can be expected to be truly alternative and where several false positive discoveries can be tolerated in exchange for more true positives. Currently, the test by Storey and Tibshirani is arguably the method of choice for such data sets. Ultimately, the choice of the test depends on the assumptions made for the data set at hand and, evidently, on further analysis of the significant features. For instance, if only very few features (say, three or four) can be included in further analysis due to financial or time constraints, then more liberal tests are of questionable benefit.

1.5.4.5 Random Permutation Tests

Random permutation tests, also known as *Monte Carlo permutation procedures*, are a special type of randomization tests. These tests use randomly generated numbers for statistical inference. Nowadays, permutation tests are being increasingly used in many practical applications, because modern standard PCs are able to cope with the computational complexity that these tests involve. Randomization tests are suitable even for very large data sets comprising numerous variables. Radmacher et al. (2002) proposed random permutation tests as an important component in building classifiers based on expression profiles. Many publications report on the application of randomization tests in the context of microarray data analysis and for testing the significance of selected marker genes, for instance, Ramaswamy et al. (2001).

To assess the significance of the weight for the ith feature, the test involves the following steps:

1. Compute the weight for the ith feature on the original data set;
2. Randomly permute the class labels;

3. Re-compute the weight for the ith feature;
4. Repeat steps (2) and (3) n times (e.g., $n = 10\,000$ times) to obtain the distribution of the weights under the null hypothesis that the feature is truly null.

The distribution under the null hypothesis can be visualized in a histogram as shown in Figure 1.6. The p-value of the ith weight on the unpermuted data set corresponds to the proportion of weights that are smaller than or equal to the observed weight.

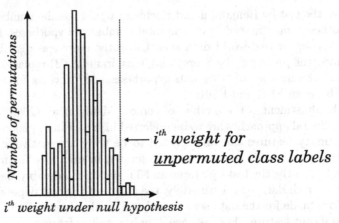

Fig. 1.6. Distribution of the weights for the ith feature under the null hypothesis.

Random permutation tests do not make any assumptions concerning the distribution of the data or the correlation structure of the features. For example, these tests do not require that the data approximate a normal distribution. The only, though important, requirement is that the random permutation experiments are independent. On the other hand, performing thousands of permutations and recomputing the weights can be computationally expensive.

1.5.5 Predictive Model Construction

After significantly different features have been selected, the analysis process often continues with the construction of a predictive model, e.g., a classifier. The basic modeling process consists of a *learning phase*, a *test phase*, and an *application phase*. The learning phase consists of the *training phase* and the *validation phase*. In the test phase, the model's performance in the analytical task at hand is ultimately assessed. The motivation for this modeling process is that one wishes a model with sufficient generalization ability. The expected prediction error of a classification model can be decomposed into two components, the *bias* and the *variance*. These two components are in a trade-off relationship, i.e., the smaller the bias, the larger the variance, and vice versa. This problem is known as the *bias-variance trade-off* in statistics, as the problem of *overfitting* in machine learning, and as *capacity control* in engineering. Essentially, this problem implies that the best generalization performance (or the smallest expected prediction error) is achieved for that model which attains the "right" balance between the accuracy on a specific learning set and the number of free model parameters. Ideally, a model should have both low variance (i.e., high precision) and low bias (i.e., high accuracy). Relatively flexible models with many free parameters tend to adapt too well to the learning data and have a relatively low bias and high variance, whereas models with fewer parameters tend to have low variance and high bias.

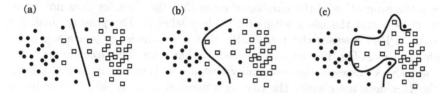

Fig. 1.7. The bias-variance trade-off in a classification problem involving two classes. Cases on the left side of the hyperplane, represented by •, are classified as members of class #1, while cases on the right side, represented by □, are classified as members of class #2. (a) A rigid (fixed) linear separator that misclassifies six cases, (b) a flexible hyperplane that misclassifies four cases, and (c) a flexible hyperplane with no misclassifications.

Figure 1.7 illustrates the bias-variance trade-off in a two-class scenario. The members of the classes are depicted by • and □, which represent a randomly selected learning set. In Figure 1.7a, a straight line separates the classes and misclassifies six cases. Assume that this line is fixed, i.e., *regardless* of the specific learning set used, exactly this line will separate the classes. This model has a large bias, but its variance is zero, since the model does not depend on the actual learning set. At the other end of the spectrum is the model shown in Figure 1.7c. This hyperplane has many parameters that allow to separate the learning data perfectly. The form of the hyperplane is highly dependent on the specific learning set, hence it has a high variance and zero bias. In other words, the model in Figure 1.7c is *overfitted.* Neither the model in Figure 1.7a nor the model in Figure 1.7c are able to generalize well. On the other hand, the model in Figure 1.7b balances the bias and the variance and is the best model in this scenario.

The higher a model's complexity, the lower its prediction error on the learning set, but the higher the error on the test set. In contrast, a model with low complexity has a high error on both the learning and the test set. A model that balances its bias and variance optimally achieves the lowest error on the test set. Essentially, the learning phase aims at finding those parameters that optimize the model's ability to generalize to unseen test cases. It is of crucial importance that the model does not adapt itself too much to the learning data ("learning by rote"), because this would result in overfitting.

1.5.5.1 Basic Measures of Performance

The most commonly used quantitative criteria are the *classification accuracy* (i.e., proportion of correctly classified cases), and, alternatively, the *error rate* (i.e., proportion of incorrectly classified cases). *Sensitivity* and *specificity* are closely related accuracy-based measures. The sensitivity of a classifier for a class y is the fraction of the number of cases that the classifier assigns to class y and the cases with class label y. The specificity of a classifier for a class y is the proportion of the number of cases that the classifier does not assign to class y and the cases without the class label y. The *positive predictive value* (PPV) assesses the probability that a case belongs to class y if the classifier classifies the case as a member of that class. The *negative predictive value* (NPV) assesses the probability that a case is not a member of y if the classifier does not classify the case as a member of y. Table 1.1 provides an overview of these measures in a confusion matrix.

Table 1.1. Accuracy-based measures for assessing classification performance.

		Real class y	$\neg y$	
Prediction	y	a (*true positive*)	b (*false positive*)	$PPV = a / (a + b)$
	$\neg y$	c (*false negative*)	d (*true negative*)	$NPV = d / (c + d)$
		sensitivity= $a / (a + c)$	specificity= $d / (b + d)$	prevalence = $(a+c)/(a+b+c+d)$

Under some circumstances, the *lift* can be a more informative measure. The lift for class y is defined as the positive predictive value divided by the prevalence. The *balanced accuracy* (BACC) is the average of sensitivity and specificity and used in Chapter 9 as evaluation criterion.

In small-sample settings, single performance scores are difficult to interpret without confidence intervals for the true statistic. For example, it does make a difference whether a classifier's correct classification rate of, say, 80% is based on 100 or on 10 000 test cases. Confidence intervals for the statistic of interest are of particular importance in scenarios comprising small data sets such as microarray data. Let M denote the number of test cases, and let m denote the number of incorrectly classified test cases. The *observed error rate* is $\epsilon = m/M$. A $(1 - \alpha)$-confidence interval for the *true error rate* τ is expected to contain τ in approximately $(1 - \alpha) \times 100\%$ of the experiments, $P(\tau_l < \tau < \tau_u | \epsilon) = 1 - \alpha$, with a balanced risk on either side of the interval, i.e., $P(\tau \leq \tau_l | \epsilon) = P(\tau \geq \tau_u | \epsilon) = \frac{1}{2}\alpha$. For deriving confidence intervals in a classification scenario, it is commonly assumed that the (integer) number of errors obeys a binomial distribution (Martin and Hirschberg, 1996). This binomial distribution is usually approximated by assuming a normal distribution of ϵ with mean τ and variance $\tau(1 - \tau)/M$. Under these assumptions, an approximate 95%-confidence interval for τ can be derived as follows:

$$\tau \approx \epsilon \pm (0.5/M + z \cdot s), \tag{1.2}$$

with $s = \sqrt{(\epsilon(1 - \epsilon)/M)}$ and $z = \Phi^{-1}(1 - \frac{1}{2}\alpha)$, e.g., $z = 1.96$ for 95% confidence, with $\Phi(\cdot)$ being the standard normal cumulative distribution function.[13] Note that Equation (1.2) involves two approximations. First, we use ϵ instead of τ for estimating the variance. Second, we approximate the binomial by the normal distribution. As a rule of thumb, these approximations are acceptable if $M\epsilon(1 - \epsilon) \geq 5$ (Mitchell, 1997).

1.5.5.2 Training, Validating, and Testing

It is essential to clearly differentiate between *observed* and *true* measures of accuracy. For instance, the observed error rate ϵ (a.k.a. *sample error*) is an estimate for a model's true error rate τ on the population of interest, for instance, the set of cases that are described by an array profile similar to the investigated data set (e.g., a "population" of similar microarray studies). The true error rate constitutes an inherent property of the model and can be interpreted as the probability that the model will misclassify a single randomly drawn instance from the distribution of the population (Mitchell, 1997). Prior to building a predictive model, an upper bound of the true prediction error, τ_{max}, should be specified. Further, the maximum number of *internal*, j_{max}, and *external*, i_{max}, (cross-)validation loops need to be initialized. Figure 1.8 depicts the modeling process. (Figure 8.1 in Chapter 8, page 182, shows this process in more detail.)

From the entire data set that is available for analysis, D, a learning set, L_i, and a test set, T_i, are sampled. Next, the learning set is split into a training set, TR_{ij}, and a validation set, V_{ij}. This sampling may or may not be done in the same way as before. The model is built (or trained) on the training set, i.e., the model parameters are adjusted/fitted using the data in TR_{ij}. The model is then calibrated using the validation set V_{ij}. This calibration involves an assessment of how well the model performs on the validation data and is a safeguard against overfitting. This splitting into training and validation sets is performed j_{max} times in the internal loop. Importantly, the model's performance on V_{ij} can be fed back into the next training round. For instance,

---

[13] The term $0.5/M$ is a continuity-correction accounting for the fact that $\Phi(\cdot)$ is a continuous and the binomial a discrete function.

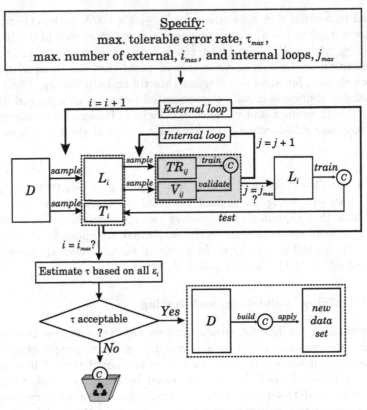

Fig. 1.8. Learning, training, validating, testing, and applying a model. D is the data set; L_i and T_i are the ith learning and test set, respectively; TR_{ij} denotes the training set of the ith external and j^{th} internal loop; V_{ij} is the corresponding validation set, and C denotes the classifier.

if the model's performance on V_{ij} is not satisfactory, then this information may be used for TR_{ij+1}. After the internal loop has been iterated j_{max} times, the "optimal" parameters are determined, e.g., the number of nearest neighbors in a nearest-neighbor classifier. Training and validation are completed for the ith learning set. The model is then built on the entire learning set L_i using these parameters and applied to the test cases in T_i. After the external loop has been iterated i_{max} times, the model's true error rate τ is estimated using the error rates ϵ_i observed in the individual external loops. If the true error rate is smaller than an arbitrarily chosen maximally tolerable error, τ_{max}, then the model is built using the entire data set, D. Those parameters are chosen that provided for the best performance on the test sets. This final model cannot be cross-validated anymore (Simon, 2003). This process is illustrated in Figure 1.9. In this example, a k-nearest neighbor classifier is to be built, with $k = 1$, $k = 3$, and $k = 5$. Both external and internal loop involve a 3-fold cross-validation in this example.

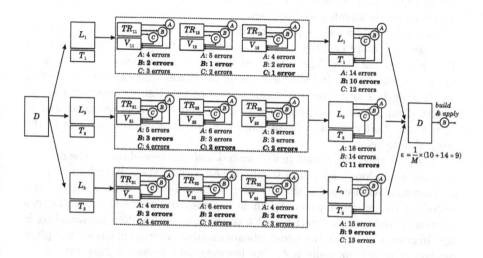

Fig. 1.9. Internal and external cross-validation loops for k-NN.

In this example, A refers to k-NN with $k = 1$, B refers to k-NN with $k = 3$, and C refers to k-NN with $k = 5$. With the exception of the second internal loop, B performed best in the internal loops. Therefore, we choose $k = 3$ and built the model using the entire data set D. We estimate the true prediction error of this model using the number of observed errors of B on the test sets T_i, i.e., $\epsilon = M^{-1}(10+14+9)$, where M is the total number of test cases. Why are the internal loops necessary in this scenario? Wouldn't it be possible to compare the 1-NN, 3-NN, and 5-NN classifiers directly using the performance on the test sets, T_i? Note that this is equivalent to comparing three distinct and fully specified classifiers, whereas the process in Figure 1.9 aims at finding the optimal parameter k for the k-NN.

It is essential to note that if feature selection is to be performed, then it must be done for each learning and training set separately to avoid a selection bias (Ambroise and McLachlan, 2002). (See Figure 9.4, Chapter 9, page 195, for examples.)

1.5.5.3 Data Resampling Strategies

There exists a variety of data resampling strategies. In *random subsampling* (a.k.a. *single hold-out method*) the cases in the data set are randomly permuted and then split into a learning set (usually, $\sim 70\%$ of the cases) and a test set ($\sim 30\%$). The classifier is constructed using the learning set and its performance is then estimated by applying the acquired classification function to the test set. This strategy is known to be suboptimal, because the classification result is highly biased by the random partitioning of the original data set into a single learning and test set.

In *k-fold random subsampling*, the described procedure is repeated k times to generate k pairs (L_i, T_i), each containing a learning set L_i and a test set T_i, with $i = 1, 2, \ldots k$. The estimated accuracy is determined by averaging the accuracies on the individual folds. It is critical that the sets L_i and T_i are disjoint, but any given two learning sets or two test sets may overlap. In *stratified* random subsampling, the learning and tests sets are generated in such a way that the class distribution is approximately the same as in the original data set.

In *k-fold cross-validation* (a.k.a. *leave-k-out cross-validation*), the original data set is randomly split into k subsets. At each of the k iterations or folds of cross-validation process, the classifier is trained on $k - 1$ data subsets and tested on the remaining subset, which becomes the test set. The difference between k-fold cross-validation and k-times repeated random subsampling is that in cross-validation, the k test sets are *disjoint*, whereas in the subsampling method they are normally not. The learning sets, however, may overlap in cross-validation. In repeated k-fold cross-validation, the described procedure is repeated n times to reduce the variance of the cross-validation error. Cross-validation can also involve a stratified sampling.

In *leave-one-out cross-validation* (LOOCV), each of the n cases of the data set is used in turn as hold-out case and the classifier is induced or trained based on the remaining $n - 1$ cases. The hold-out case is the test case. The classification accuracy can be estimated as the number of correctly classified hold-out cases divided by n. LOOCV, which is computationally expensive as it involves n times a complete model re-calibration (including feature selection) for a data set of n cases. The estimate for the prediction error is almost unbiased and therefore has been recommended for small-sample microarray data (see Chapter 8). LOOCV is one of the most frequently used methods to estimate classification performance in microarray studies (Simon, 2003).

Another approach is *bootstrapping*. Here, b subsamples are drawn from the original data set with replacement, so that b bootstrap data sets are available to estimate a statistic. Kohavi (1995) presents examples where LOOCV and the .632 bootstrap method (Efron and Tibshirani, 1993) fail to provide reliable estimates for classification accuracy and recommends stratified 10-fold cross-validation. Braga-Neto and Dougherty (2004) reported that they were not able to verify a substantial difference in performance between bootstrap and various cross-validation approaches for microarray data. Experimental evidence shows that repeated stratified 10-fold cross-validation is an appropriate method for assessing classification performance (Bouckaert and Frank, 2004), but its application to microarray data analysis is problematic due to the small sample size (Li et al., 2004; Braga-Neto and Dougherty, 2004).

Generally, the larger the size of the learning set, the more effective will the classifier be in terms of its generalization ability. In contrast, the larger the test set, the more reliable is the error estimate obtained from applying the classifier to the test set. Chapter 8 discusses resampling strategies in more detail.

1.5.6 Statistical Significance Tests for Comparing Models

Many studies involve comparisons of models for classification. For instance, if a novel classifier is developed, then it is important to benchmark its performance against established models. It is common practice to compare model performance by referencing published reports on competing models. However, cross-study comparisons of monolithic accuracy-based measures of performance are difficult to interpret. For example, suppose that a classifier A achieved an accuracy rate of $x\%$ on a particular data set D, whereas a classifier B achieved $y\%$ on the same data set. The observed difference, $|x\% - y\%|$, in accuracy could, for instance, have been caused by different experimental settings (e.g., different sampling strategies), rather than by inherent differences in the analytical classification methods. Comparing monolithic accuracy measures or (overlapping) confidence intervals for error rates is not appropriate to compare models. Essentially, the key question is whether the observed differences in performance provide sufficient evidence to conclude that the models perform significantly differently, or whether we cannot exclude the possibility (with reasonable confidence) that this difference may be due to chance alone or to the random variation introduced by the sampling strategy (Dietterich, 1998). The following three aspects need to be taken into account when models are compared (Berrar et al., 2006):

1. The learning and test sets should be identical for the classifiers;
2. The learning phases should include a complete parameter re-calibration and external cross-validation;
3. The difference in performance should be assessed by means of a suitable statistical test, which is appropriate for the adopted sampling strategy and accounts for both comparison- and family-wise error rates by adjusting for multiple testing.

It is essential that the learning and test sets are identical for the classifiers. Otherwise, observed differences in performance could be due to differences in the make-up of the sampled data sets. The learning phases should include a complete re-calibration of the model's parameters, for instance, the number of nearest neighbors in the k-NN classifier. Finally, the differences in performance need to be assessed by means of statistical tests. If a comparative study includes more than two classifiers, then it is also necessary to correct for multiple hypotheses testing. And consequently, it is also advisable to carefully select the competing models, because by including too many classifiers, it is more difficult to control the family-wise error rate at the desired level. Clearly, different classifiers have different strengths and weaknesses. For instance, it is known that support vector machines perform remarkably well in binary classification tasks. Therefore, it is necessary that novel models are benchmarked against the most similar established models (Salzberg, 1997).

If the performance of two models is compared on one single test set, then *McNemar's test* can be applied (Ripley, 1996). Under the null hypothesis that two classifiers A and B perform equally, the respective error rates should be the same. McNemar's test is based on a χ^2 test for goodness-of-fit that compares the distribution of errors expected under the null hypothesis of equal performance to the observed errors. The test statistic X is distributed approximately as χ^2 with 1 degree of freedom and incorporates a continuity correction term of -1 in the nominator to account for the fact that the statistic is discrete whereas the χ^2 distribution is continuous. Let m_A be the number of test errors made by model A but not by model B, and let m_B be the number of test errors made by model B but not by A.

$$X = \frac{(|m_A - m_B| - 1)^2}{m_A + m_B} \tag{1.3}$$

If the null hypothesis is correct, then the probability that X is greater than 3.84 is less than 0.05. However, this test cannot be applied if the sampling strategy is LOOCV, because the binary counts of misclassifications are not independent from each other due to the overlapping learning sets. Furthermore, it is common practice to evaluate the performance using multiple test sets. Dietterich (1998) recommends deriving a t-statistic based on five replications of 2-fold cross-validation (5×2CV test). Bouckaert and Frank (2004) compared various tests and recommended a variance-corrected resampled t-statistic.

Let p_{A_i} be the observed proportion of test cases misclassified by model A and let p_{B_i} be the observed proportion of misclassified test cases by B during the ith fold. If we assume that the differences $p_i = p_{A_i} - p_{B_i}$ were drawn independently from a normal distribution, then we could apply Student's t-test. However, the assumptions underlying this test are violated, because in cross-validation and repeated random subsampling, the learning sets necessarily overlap. In repeated random subsampling, the test sets usually overlap as well. Hence, the individual differences p_i are not independent from each other. The high Type I error of Student's t-test is due to an underestimation of the variance because the samples are not independent.

Let the number of folds be k, and let the number of repetitions be r. In each fold and in each repetition, the number of learning cases is N and the number of test cases is M. The proportion of cases that A misclassifies is $p_{A_{ij}} = m_{A_{ij}}/M$, with $m_{A_{ij}}$ the number of errors on the ith test set in the jth repetition (analogously for $p_{B_{ij}}$ and $m_{B_{ij}}$). The difference of proportion of misclassified cases in the ith fold of the jth repetition is $p_{ij} = p_{A_{ij}} - p_{B_{ij}}$. The average difference is then $\bar{p} = \frac{1}{rk} \sum_{i=1}^{r} \sum_{j=1}^{k} p_{ij}$ and the estimated variance of the r times k differences is $\bar{p} = \frac{1}{rk-1} \sum_{i=1}^{r} \sum_{j=1}^{k} (p_{ij} - \bar{p})^2$. The statistic for the variance-corrected resampled paired t-test for r times repeated k-fold cross-validation is then given by Equation (1.4) (Nadeau and Bengio, 2003).

$$T = \frac{\bar{p}}{\sqrt{(\frac{1}{rk} + \frac{M}{N})s^2}} \qquad (1.4)$$

The learning set should be at least five times larger than the test set. The statistic follows approximately Student's t-distribution with $rk - 1$ degrees of freedom. Empirical results show that this corrected statistic drastically improves on the standard resampled t-test with respect to the Type I error (Nadeau and Bengio, 2003; Bouckaert and Frank, 2004). This statistic can also be used for repeated random subsampling.

1.6 Result Post-Processing

Statistical significance does not necessarily imply biological relevance. It is possible that microarray data sets comprise thousands of potential marker genes. While we can construct classifiers for such a number of predictors, the investigation of the biological relevance of all these markers is nearly impossible due to time, financial, and other constraints. The life scientist faces the challenge of selecting a manageable set of predictor variables (e.g., genes) for further investigation. However, a simple ranking of the individual predictors based on the obtained p-values is generally not what the life scientist is interested in. For instance, those genes that score the smallest p-values for a cancer type might not be of interest, because they are either obvious or irrelevant. In cancer genomics, for example, a large number of genes found to be overexpressed in tumor cells simply reflect the fact that aggressive cancer cells tend to be in active cell cycle, and thus express genes that are known to be expressed in cycling cells (O'Neill et al., 2003). Consequently, it is possible that some genes with a larger p-value are of greater biological relevance. Usually, the scientist is not primarily interested in the "top-scoring" features, but in a set of features that, in addition to being significant, exhibit "interesting" characteristics.

1.6.1 Statistical Validation

It is essential that the study is conducted within a stringent statistical framework as outlined above; for instance, correction for multiplicity effects is of paramount importance. But statistical validation does not necessarily end here. It is important to demonstrate that the results are reproducible between samples, experimenters, laboratories, and dates of experiments (Baggerly et al., 2004; Simon, 2005). Clinical drug trials generally follow the Good Clinical Practice guidelines and are prospective with respect to patient selection and primary study endpoints. In contrast, exploratory genomic and proteomic studies are characterized by multiple endpoints and multiple hypotheses, which increases the chance that experimental and study-specific artifacts are identified as biological phenomena. Consequently, before a classifier is to be applied in a clinical setting, it is indispensable to demonstrate its therapeutic relevance by a prospectively planned validation study (Simon, 2005).

1.6.2 Epistemological Validation

Most techniques that are currently applied in mining high-throughput genomic and proteomic data mainly focus on statistical and machine learning approaches, and fail to incorporate the huge body of formal background knowledge that already exists in the form of scientific articles, databases, and gene ontologies. Valuable gold nuggets of information remain to be mined out of these resources. Text mining might become the tool for digging out those hidden nuggets. For instance, purely number-based approaches are arguably not sufficient to uncover insights into those genes that might be causal for the development or progression of a phenotype. Here, text mining approaches could help identify those genes that share the same functionality, or whose products share a similar cellular localization. Chapter 12 discusses text mining in the biological context.

1.6.3 Biological Validation

Validating the experimental and analytical results by means of other biotechnological techniques represents the final step in the analysis process. Findings from high-throughput genomics and proteomics studies are still primarily of scientific – not therapeutic – interest. An individualized, patient-tailored drug dosage based on genomic profiles would constitute an outstanding achievement.

1.7 Conclusions

A frequently asked question to the data analyst is: "*Which is the best method for ... ?*" The answer might be sobering. There is no free lunch for anyone – every statistical approach is as good as the assumptions that you impose on them. For example, the "best" approach for correction of multiple testing depends on how liberal or conservative you decide to be. Resampling strategies are indispensable in studies that are characterized by the small-n-large-p problem. All resampling strategies and statistical tests for assessing model performance in resampled data sets, however, have their intrinsic problems (Kohavi, 1996; Mitchell, 1997; Dietterich, 1998).

   Which classifier is the method of choice? This question is arguably ill-posed, since the success of a specific method ultimately depends on the specific data set at hand. Support vector machines have been successfully employed in many genomics and proteomics studies and shown excellent performance in comparative studies (Li et al., 2004; Statnikov et al., 2005). Nevertheless, in theory, SVMs are not inherently able to overcome the curse of dimensionality (Hastie et al., 2002), and often the performance of simpler methods (e.g., nearest neighbor classifiers) is comparable to that of SVMs (Dudoit et al., 2002). From an Occam's razor perspective, it is advisable to consider simpler classifiers for genomics and proteomics data sets (Simon, 2005).

Data pre-processing does not mean that the data should be tortured until they confess. As a general rule of thumb, the analyst should pre-process the data as little as possible and as much as necessary.

Finally, it is advisable to include a statistician prior to carrying out the experiment – "*To call in the statistician after the experiment is done may be no more than asking him to perform a postmortem examination: He may be able to say what the experiment died of.*" (Sir R. Fisher, 1890–1962).

References

Alter, O., Brown, P.O., and Botstein, D. (2000). Singular-value decomposition for genome-wide expression data processing and modeling. *Proc. Natl. Acad. Sci. USA*, 97(18):10101–10106.

Ambroise, C. and McLachlan, G.J. (2002). Selection bias in gene extraction on th basis of microarray gene expression data. *Proc. Natl. Acad. Sci. USA*, 98:6562–6566.

Baggerly, K.A., Morris, J.S., and Coombes, K.R. (2004). Reproducibility of SELDI-TOF protein patterns in serum: Comparing datasets from different experiments. *Bioinformatics*, 20(5):777–785.

Bartlett, M.S. (1937). Properties of sufficiency and statistical tests. *Proc. R. Stat. Soc. Series A*, 160:268–282.

Benjamini, Y. and Hochberg, Y. (1995). Controlling the false discovery rate: A practical and powerful approach to multiple testing. *J. Roy. Stat. Soc.*, B57:289–300.

Berrar, D., Bradbury, I., and Dubitzky, W. (2006). Avoiding model selection bias in small-sample genomic data sets. *Bioinformatics*, 22(10):1245–1250.

Berry, M.J.A. and Linoff, G. (1997). *Data Mining Techniques: For Marketing, Sales, and Customer Support.* Wiley, USA.

Bouckaert, R.R. and Frank, E. (2004). Evaluating the replicability of significance tests for comparing learning algorithms. *Proc. 8th Pacific-Asia Conference on Knowledge Discovery and Data Mining*, 3056:3–12.

Braga-Neto, U.M. and Dougherty, E. (2004). Is cross-validation valid for small-sample microarray classification? *Bioinformatics*, 20(3):374–380.

Brown, M.B. and Forsythe, A.B. (1974). Robust tests for the equality of variances. *J. Am. Stat. Ass.*, 69:264–267.

Burnette, N.W. (1981). "Western Blotting": Electrophoretic transfer of protein sodium dodecyl sulfate-polyacrylamid gels to unmodified nitrocellulose and radiographic detection with antibody and readiojodinated protein. *Anal. Biochem.*, 112:195–203.

Bustin, S.A. (2000). Absolute quantification of mrna using real-time reverse transcription polymerase chain reaction assays. *J. Mol. Endocrinol.*, 25:169–193.

Chen, D., Liu, Z., Ma, X., and Hua, D. (2005). Selecting genes by test statistics. *J. Biomed. Biotech.*, 2:132–138.

Cochran, W.G. (1937). Problems arising in the analysis of a series of similar experiments. *J. Roy. Stat. Soc. Ser. C. Appl. Stat.*, 4:102–118.

Diatchenko, L., Lau, Y.F., and Campbell A.P., et al. (1996). Suppression subtractive hybridization: a method for generating differentially regulated or tissue-specific cDNA probes and libraries. *Proc. Natl. Acad. Sci. USA*, 93(12):6025–6030.

Dietterich, T. (1998). Approximate statistical tests for comparing supervised classification learning algorithms. *Neural Comp.*, 10(7):1895–1924.

Dudoit, S., Fridlyand, J., and Speed, T.P. (2002). Comparison of discrimination methods for the classification of tumors using gene expression data. *J. Am. Stat. Assoc.*, 97:77–87.

Efron, B. and Tibshirani, R. (1993). *An Introduction to the Bootstrap*. Chapman & Hall.

Fields, S. and Song, O. (1989). A novel genetic system to detect protein-protein interactions. *Nature*, 340:245–246.

Glish, G.L. and Vachet, R.W. (2003). The basics of mass spectrometry in the twenty-first century. *Nat. Rev. Drug Discov.*, 2(2):140–150.

Golub, T.R., Slonim, D.K., and Tamayo P., et al. (1999). Molecular classification of cancer class discovery and class prediction by gene expression monitoring. *Science*, 286(5439):531–537.

Hastie, T., Tibshirani, R., and Friedman, J. (2002). *The Elements of Statistical Learning*. Springer Series in Statistics, New York/Berlin/Heidelberg.

Hedenfalk, I., Ringnér, M., Ben-Dor, A., Yakhini, Z., Chen, Y., Chebil, G., Ach, R., Loman, N., Olsson, H., Meltzer, P., Borg, A., and Trent, J. (2003). Molecular classification of familial non-BRCA1/BRCA2 breast cancer. *Proc. Natl. Acad. Sci. USA*, 100(5):2532–2537.

Hochberg, Y. (1988). A sharper Bonferroni procedure for multiple tests of significance. *Biometrika*, 75:800–802.

Hod, Y. (1992). A simplified ribonuclease protection assay. *Biotechniques*, 13:852–854.

Holm, S. (1979). A simple sequentially rejective multiple test procedure. *Scand. J. Stat.*, 6:65–70.

Honoré, B., Ostergaard, M., and Vorum, H. (2004). Functional genomics studied by proteomics. *Bioessays*, 26(8):901–915.

Hoogenboom, H.R., de Bruine, A.P., Hufton, S.E., Hoet, R.M., Arends, J.W., and Roovers, R.C. (1998). Antibody phage display technology and its applications. *Immunotechnology*, 4(1):1–20.

Issaq, H.J., Veenstra, T.D., Conrads, T.P., and Felschow, D. (2002). The SELDI-TOF MS approach to proteomics: Protein profiling and biomarker identification. *Biochem. Biophys. Res. Commun.*, 292(3):587–592.

Johansson, P. and Hakkinen, J. (2006). Improving missing value imputation of microarray data by using spot quality weights. *BMC Bioinformatics*, 7(1):306.

Karas, M., Bachmann, D., Bahr, U., and Hillenkamp, F. (1987). Matrix-assisted ultraviolet laser desorption of non-volatile compounds. *Int. J. Mass Spectrom. Ion Processes*, 78:53–68.

Klipp, E., Herwig, R., Kowald, A., Wierling, C., and Lehrach, H. (2005). *Systems Biology in Practice*. Wiley-VCH, Weinheim, Germany.

Klose, J. and Kobalz, U. (1995). Two-dimensional electrophoresis of proteins: An updated protocol and implications for a functional analysis of the genome. *Electrophoresis*, 16(6):1034–1059.

Kohavi, R. (1995). A study of cross-validation and bootstrap for accuracy estimation and model selection. *Proc. 14th Intl. Joint Conf. Art. Int.*, pages 1137–1143.

Kruskal, W.H. and Wallis, W.A. (1952). Use of ranks in one-criterion variance analysis. *J. Am. Stat. Ass.*, 47:583–621.

Levene, H. (1960). Robust tests for equality of variances. *Contributions to Probability and Statistics*, pages 278–292.

Li, T., Zhang, C., and Ogihara, M. (2004). A comparative study of feature selection and multiclass classification methods for tissue classification based on gene expression. *Bioinformatics*, 20(15):2429–2437.

Liang, P. and Pardee, A.B. (1992). Differential display of eukaryotic messenger RNA by means of the polymerase chain reaction. *Science*, 257(5072):967–971.

Lorkowski, S. and Cullen, P. (2003). *Analysing Gene Expression: A Handbook of Methods Possibilities and Pitfalls*. Wiley-VCH, Weinheim, Germany.

MacBeath, G. (2002). Protein microarrays and proteomics. *The Chipping Forecast II, Nat. Gen.*, 32:526–532.

Manly, K.F., Nettleton, D., and Hwang, J.T.G (2004). Genomics, prior probability, and statistical tests of multiple hypotheses. *Genome Res.*, 14:997–1001.

Martin, J.K. and Hirschberg, D.S. (1996). Small sample statistics for classification error rates II: Confidence intervals and significance tests. Technical Report 96-22, University of California, Irvine, CA.

Mitchell, T.M. (1997). *Machine Learning*. McGraw-Hill Book Co., Singapore.

Moody, D.E. (2001). Genomics techniques: An overview of methods for the study of gene expression. *J. Anim. Sci.*, 79(E.Suppl.):E128–135.

Morris, J.S., Yin, G., Baggerly, K., Wu, C., and Zhang, L. (2003). Identification of prognostic genes, combining information across different institutions and oligonucleotide arrays. *Oral and Poster Presenters' Abstracts, 4th Int. Conf. Critical Assessment of Methods for Microarray Data Analysis*, pages 1–5.

Morrison, N. and Hoyle, D.C. (2002). Normalization – Concepts and methods for normalizing microarray data. In Berrar, D., Dubitzky, W., and Granzow, M., editors, *A Practical Approach to Microarray Analysis*, pages 76–90. Kluwer Academic Publisher, Boston.

Murphy, D. (2002). Gene expression studies using microarrays: Principles, problems, and prospects. *Adv. Physiol. Educ.*, 26(4):256–270.

Nadeau, C. and Bengio, Y. (2003). Inference for generalization error. *Machine Learning*, 52:239–281.

O'Farrell, P.H. (1975). High-resolution two-dimensional gel electrophoresis of proteins. *J. Biol. Chem.*, 250(10):4007–4021.

O'Neill, G.M., Catchpoole, D.R., and Golemis, E.A. (2003). From correlation to causality: Microarrays, cancer, and cancer treatment. *BioTechniques*, 34:S64–S71.

Radmacher, M.D., McShane, L.M., and Simon, R. (2002). A paradigm for class prediction using gene expression profiles. *J. Comp. Bio.*, 9(3):505–511.

Ramaswamy, S., Tamayo, P., and Rifkin, R., et al. (2001). Multiclass cancer diagnosis using tumor gene expression signatures. *Proc. Natl. Acad. Sci. USA*, 98(26):15149–15154.

Raychaudhuri, S., Stuart, J.M, and Altman, R.B. (2000). Principal components analysis to summarize microarray experiments: Application to sporulation time series. *Proc. 5th Pac. Symp. Biocomp.*, pages 455–566.

Ripley, B.D. (1996). *Pattern Recognition and Neural Networks*. University Press, Cambridge.

Saiki, R.K., Gelfand, D.H., Stoffel, S., Scharf, S.J., Higuchi, R., Horn, G.T., Mullis, K.B., and Erlich, H.A. (1988). Primer-directed enzymatic amplification of DNA with a thermostable DNA polymerase. *Science*, 239(4839):487–491.

Salzberg, S. (1997). On comparing classifiers: Pitfalls to avoid and a recommended approach. *Data Mining and Knowledge Discovery*, 1:317–327.

Sargent, T.D. and Dawid, I.B. (1983). Differential gene expression in the gastrula of *xenopus laevis*. *Science*, 222(4620):135–139.

Schena, M., Shalon, D., Davis, R.W., and Brown, P.O. (1995). Quantitative monitoring of gene expression patterns with a complementary DNA microarray. *Science*, 270(5235):467–470.

Simon, R. (2002). Classifying breast cancer models. *The Scientist*, 16(17).

Simon, R. (2003). Supervised analysis when the number of candidate features (*p*) greatly exceeds the number of cases (*n*). *SIGKDD Explorations*, 5(2):31–36.

Simon, R. (2005). Roadmap for developing and validation therapeutically relevant genomic classifiers. *J. Clin. Onc.*, 23(29):7332–7341.

Somogyi, R., Fuhrman, S., and Wen, X. (2002). Genetic network inference in computational models and applications to large-scale gene expression data. In Bower, J.M. and Bolouri, H., editors, *Computational Modeling of Genetic and Biochemical Networks*, pages 119–157.

Somorjai, R.L., Dolenko, B., and Baumgartner, R. (2003). Class prediction and discovery using gene microarray and proteomics mass spectroscopy data: Curses, caveats, cautions. *Bioinformatics*, 19(12):1484–1491.

Statnikov, A., Aliferis, C.F., Tsamardinos, I., Hardin, D., and Levy, S. (2005). A comprehensive evaluation of multicategory classification methods for microarray gene expression cancer diagnosis. *Bioinformatics*, 21(5):631–643.

Storey, J.D. and Tibshirani, R. (2003). Statistical significance for genomewide studies. *Proc. Natl. Acad. Sci. USA*, 100(16):9440–9445.

Tang, N., Tornatore, P., and Weinberger, S.R. (2004). Current developments in SELDI affinity technology. *Mass. Spectrom. Rev.*, 23(1):34–44.

Troyanskaya, O., Cantor, M., Sherlock, G., Brown, P., Hastie, T., Tibshirani, R., Botstein, D., and Altman, R.B. (2001). Missing value estimation methods for DNA microarrays. *Bioinformatics*, 17(6):520–525.

Unlu, M., Morgan, M.E., and Minden, J.S. (1997). Difference gel electrophoresis: A single gel method for detecting changes in protein extracts. *Electrophoresis*, 18(11):2071–2077.

Velculescu, V.E., Zhang, L., Vogelstein, B., and Kinzler, K.W. (1995). Serial analysis of gene expression. *Science*, 270(5235):484–487.

Welch, B.L. (1951). On the comparison of several mean values: An alternative approach. *Biometrika*, 38:330–336.

Wolpert, D. and Macready, W. (1997). No free lunch theorems for optimization. *IEEE Trans. Evolut. Comp.*, 1(1):67–82.

Yamashita, M. and Fenn, J.B. (1984). Electrospray ion source, another variation of the free-jet theme. *J. Phys. Chem.*, 88:4451–4459.

Design Principles for Microarray Investigations

Kathleen F. Kerr

Department of Biostatistics, University of Washington, Seattle, Washington 98195, USA.
katiek@u.washington.edu

2.1 Introduction

In the past decade, high-throughput measurement of gene expression has evolved from a tantalizing possibility to an everyday exercise, thanks to microarray technology. The initial excitement for microarrays was quickly followed, for many scientists, with apprehension about appropriately analyzing large amounts of data of sometimes questionable quality. Most scientists have now developed an appreciation for the limitations and challenges presented by the technology.

A microarray study should not be conducted without careful thought and planning, even if it is exploratory. As with any other type of scientific investigation, a successful microarray study starts with developing a well-defined project with well-defined goals. One must then develop and implement a sound experimental design based on these goals. This chapter will begin with a discussion of some of the basic issues to consider in the earliest stages of planning a microarray study. In Section 2.3, I discuss three general principles of statistical design that apply generally to scientific experimentation: *Replication*, *blocking*, and *randomization*. We will review each of these concepts in turn, and discuss each of them in the context of array experiments.

2.2 The "Pre-Planning" Stage

By the time a scientist consults with a statistician about the experimental design for a microarray study, she has probably already made some important design choices. The scientist has probably already chosen the types of mRNA to be studied. That is, she has chosen the organism and tissue type, and has decided which treatments to apply or under what conditions the mRNA will be collected. These choices are primarily made based on scientific, not statistical, considerations, although a technical consideration is whether the samples can provide a sufficient amount of mRNA for the assay.

At this stage, it is important to recognize whether a study is an *experiment* or an *observational study*. Unfortunately, microarray studies all tend to be called "experiments," but this can be a misnomer (Potter, 2003). For example, consider a study in which tissue samples are compared between patients with a particular kind of cancer and cancer-free control subjects. The investigator does not assign cancer status to the subjects, he is merely making measurements on a sample of cases and controls. This is an observational study, even though the observations happen to be measurements of gene expression for thousands of genes. The fact that the investigation is an observational study has profound implications for the interpretation of the data. For example, the investigator would *not* be automatically justified in attributing any observed differences in gene expression between the cases and controls to their cancer status because the differences could be due to a *confounding factor*. That is, the cases and controls might differ in their distributions of age, sex, environmental exposures, or what they ate for breakfast. Unfortunately, in many such observational microarray studies, data on potential confounding factors are not collected and the possible impact of such factors is ignored. Such gross oversight makes an entire study scientifically questionable (Potter, 2003).

In the early planning stage, it is important to establish realistic expectations for the array study. Because arrays produce more data than many biologists are used to, some biologists make the natural leap that they produce a vast amount of information. In a sense they do, but the information is far from complete and a successful array study will produce at least as many questions as it answers. Thus, it is important to clarify the goals of the array experiment. Dudoit et al. (2002) describe three distinct goals of microarray experiments: *Unsupervised learning* (Goal 1), *supervised learning* (Goal 2), and *class comparison* (Goal 3). I discuss each of these briefly, then focus on Goal 3 for the remainder of this chapter.

2.2.1 Goal 1: Unsupervised Learning

In very general terms, unsupervised learning attempts to organize data into groups of "similar" observations. With microarray data, this might mean using gene expression data on multiple genes to organize or "cluster" subjects into groups with similar gene expression profiles. Alternatively, one could organize genes into groups within which the expression profiles are similar across individuals. Eisen et al. (1998) presented an early and influential microarray paper that demonstrated the application of a particular flavor of unsupervised learning called hierarchical clustering. Sometimes clustering subjects and clustering genes are done simultaneously; this is especially common when hierarchical clustering is used. See Chapter 6 of this book for more information on unsupervised learning techniques. Note that unsupervised learning is also called *class discovery* and, most often in microarrays, *cluster analysis*.

Sometimes unsupervised learning is used with a specific goal in mind, for example, discovering new sub-types of cancer that have previously been hypothesized to exist. More commonly, unsupervised learning is used as a completely exploratory technique. There is an emerging consensus that unsupervised techniques are overused (Allison et al., 2006), as many studies that use these techniques would be better served supervised learning (Section 2.2.2) or class comparison (Section 2.2.3) approaches.

The literature contains little discussion of design issues for studies in which unsupervised learning will be used. Dobbin and Simon (2002) may be the only paper on the subject. However, the lack of research in this area should not be interpreted as an indication that design issues are not important in these studies. Section 2.3.3 of this chapter gives an example that illustrates how poor design can produce misleading results in cluster analysis.

2.2.2 Goal 2: Supervised Learning

Supervised learning is also know as supervised classification and discriminant analysis. An example application is a study where the goal is to develop an algorithm to make an accurate prognosis for cancer patients based on gene expression measurements on biopsy samples. An accurate prognosis could help patients and their doctors decide whether to pursue more aggressive treatment. The data include information on the eventual outcome for the subjects, and this information is used to develop (or "train") the algorithm, which is why the learning is called "supervised." See Chapter 9 for more information on supervised learning techniques.

Supervised learning is typically done with the possibility of a clinical application in mind. As such, the data used in a supervised learning analysis are invariably from an observational study, not an experiment. A truly useful classification algorithm must be able to classify new subjects, not just those in the sample. An important factor for facilitating this is to ensure that there are no obvious differences between the kinds of samples in study design. For example, suppose the biopsy samples for long-term cancer survivors tend to be older, whereas the samples for patients who died quickly tend to be fresher. Handling and storage differences could affect the array measurements, and these differences could influence the parameters of the classification algorithm. Thus, an algorithm that putatively discriminates between patients with good and poor prognoses is actually distinguishing between handling and storage differences between the RNA. Because of this design flaw, the algorithm will not perform well when tested on new samples from newly-diagnosed patients, all of whom provide fresh samples.

2.2.3 Goal 3: Class Comparison

Class comparison is probably the most common goal of gene expression studies and is the focus of the remainder of this chapter. In a typical class comparison study, an investigator wants to identify genes that are differentially expressed between two or more classes of tissue. A class comparison investigation can be either an experiment or an observational study. For example, a comparison between laboratory mice treated with a certain drug and untreated mice is an experiment, as long as the pre-specified number of mice to receive the treatment are chosen randomly from all mice in the study. In contrast, a study that identified differentially expressed genes between patients with and without a particular malignancy is an observational study.

In class comparison studies it is important to understand that microarrays do not remove inherent limitations in determining the "cause and effect" in some system. As a measurement tool, microarrays cannot be used to make causal inferences unless the study is explicitly designed to make this possible. In the observational study comparing malignant tissue with benign controls, microarrays cannot distinguish genes whose altered expression *caused* the malignancy from genes whose expression is altered *as a result of* the malignancy. In fact, the study can only conclude that altered expression is *associated with* the malignancy, keeping in mind that such an association could be due to a confounding factor (Potter, 2003).

In the microarray experiment with the treated and untreated mice, we can justify causal inference about the effect of the drug on gene expression because of the initial randomization of the treatment. However, note that the causal inference is about the effect of the treatment. This is quite different from trying to infer the causal effect of gene expression changes.

Once these basic issues have been considered, the next step is to plan the details of the microarray study itself. We now discuss the three fundamental principles of design, replication, blocking, and randomization, focusing on their application to microarrays and in particular to microarray studies for class comparison.

2.3 Statistical Design Principles, Applied to Microarrays

2.3.1 Replication

Replication is probably the most widely-recognized principle of design. Researchers carefully plan the sample size of their studies to ensure adequate replication.

To appreciate the important role of replication, it is useful to review the general paradigm of statistics. Scientifically, we are often interested in comparing different groups or classes of individuals: Treated and untreated; diseased and non-diseased; genotypes AA, Aa, and aa (see class comparison, Section 2.2.3). In statistics, such groups are called *populations*. A population is generally either very large or infinite, so it is impossible to examine an entire population. Instead, we take a *sample* from the population. We may study the sample in excruciating detail, collecting and analyzing data. Ironically, however, our true interest is not in the individuals in the sample. Our interest in the sample is as a means to making *inference* to the population from which it was drawn. A statistical inference is something more than a generalization or an educated guess. The theory of statistics allows us to make inferences with rigor: Using the data on a random sample, we can estimate certain characteristics of a population (for example, the mean expression of gene xyz in the population), and we can also quantify our level of certainty in the estimate (often, with a confidence interval). However, rigorous statistical inference is only possible with replication. In other words, samples of size 1 are not sufficient. Further, an adequate level of precision in inference is achieved only with an adequate amount of replication.

Understanding this fundamental statistical paradigm can help a researcher understand the appropriate level on which to replicate. In research with microarrays, it is common to differentiate between *technical replicates* and *biological replicates* (Yang and Speed, 2002). Technical replicates are typically repeated hybridizations of the same RNA to multiple arrays. Replication in early array experiments was often limited to technical replication. Technical replication allows one to make inference about the particular RNAs being studied in light of the technical error (measurement error) of the assay. However, this is usually not the desired inference. Most often, the desired inference is from the sampled individuals to the population(s) they represent. This inference is only possible with biological replication: Multiple individuals sampled from each population of interest.

Kerr (2003a) examines the relative benefits of biological and technical replication. Technical replication can be useful, but is usually unnecessary. It is usually best to use available resources to maximize biological replication and forego technical variation altogether (Simon et al., 2002; Kerr, 2003a).

2.3.2 Blocking

The term "blocking" comes from the agricultural origins of the field of statistical design. Suppose one wants to conduct a study to compare, say, the yields of different varieties of a crop. Suppose further that different blocks of land are available to use in the study. Different blocks of land will vary in many characteristics that can affect yield, e.g., the amount of sunlight or the soil composition. It would be crucial to recognize this in planning the experiment. The more variation among the blocks of land, the more important it is to explicitly address this source of variation in the experimental design. If block-to-block variability is large, an effective solution is to balance varieties with respect to blocks. For example, if there are four varieties and each block can accommodate four sub-plots, then each block should contain one of each variety (Figure 2.1). In statistical design this would be called a "complete block design." "Complete" refers to the fact that every block contains an equal number of replicates of each variety.

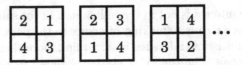

Fig. 2.1. An experiment in which the experimental units come in blocks of size 4. If there are four groups to compare, the best design is to put one of each variety in each block.

Experimentalists routinely and intuitively use the principle of blocking. For example, if an assay is known to be sensitive to humidity, then an experimentalist may make sure to conduct all assays within a short period of time when humidity is constant. Two ocular treatments might be compared by applying each of them to one eye of multiple individuals. Each pair of eyes is a "block" in such a study design. This design controls for variation between individuals by enabling the treatments to be compared "within" each individual.

In microarray studies, it can be important and useful to implement blocking as with any other kind of experiment. For example, if treatments are to be compared on mice from various litters, a litter of mice should be treated as a block. Ideally, each treatment could be applied to the same number of mice in each litter.

For two-color microarray platforms, blocking is intrinsic to the technology. This is because spot characteristics (size, density, etc.) are variable, which means a large signal could result from a high level of gene expression or from a particularly large or dense spot. However, if spot characteristics lead to a high level of signal, then the signal should be brighter in *both* channels. Therefore, the *relative* sizes of the red and green signals is used as a measure of the *relative* levels of expression in the red- and green-labeled RNAs. In other words, ratios are used because they control for spot-to-spot variation from array to array. Taking ratios (or better, log-ratios) "cancels out" uninteresting variation that is due to spot heterogeneity. This is actually a textbook example of the principle of blocking.

While the majority of analyses are based on the ratio of the red and green signals from each spot, some analytical methods start with the individual signal intensities rather than ratios. For example, see Kerr et al. (2000) and Wolfinger et al. (2001). Such methods simply handle the blocking structure of the data in a different way. In fact, the difference between intensity-based methods and ratio-based methods is somewhat more technical than substantive – see (Kerr, 2003b).

Because of spot heterogeneity, two-color arrays are used to measure relative gene expression, not absolute gene expression. A two-color array can be thought of as a comparison between the co-hybridized RNAs. When there are multiple samples to be compared, this raises the question: Which hybridizations to perform? That is, what pairs of RNAs should be co-hybridized? Kerr and Churchill (2001) addressed this question for experiments that do not contain biological replicates. Dobbin and Simon (2002) and Kerr (2003a) update these findings for experiments with biological replicates.

When there are n replicates from two groups to be compared, an efficient and effective strategy is the multiple-dye-swap design, as seen in Figure 2.2(a). In this design, the n replicates from the two groups are randomly paired and each pair is co-hybridized to a pair of arrays, with a dye-swap to control for dye-effects. Another design, similar to those proposed by Rosa et al. (2005), is to alternate the dye-labeling between replicates (see Figure 2.2(b)). This will allow twice the number of replicates to be used for the same cost of arrays, while maintaining dye-balance. Another, popular strategy is to employ a "reference" RNA in the design; each RNA of interest is co-hybridized with the reference RNA. The reference RNA is not of interest and serves only to "connect" the other samples. In Figure 2.2(c), this strategy is employed for the two-group comparison problem, employing dye-swap. While the reference design is technically less efficient than the multiple-dye swap strategy, its efficiency disadvantage is small when biological variation is much larger than technical variation (Kerr, 2003a). It is an exceedingly simple and practical design choice for many investigations.

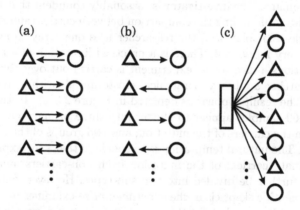

Fig. 2.2. Circles represent biological replicates from some population and triangles represent biological replicates from another population. Arrows represent two-color microarrays. An arrow between individual 1 and individual 2 indicates a hybridization with red-labeled RNA from individual 1 and green-labeled RNA from individual 2. All designs are appropriate for a two-group comparison study. (a) Multiple dye-swap design; (b) Alternating-dye pairwise design; (c) Reference design – the rectangle represents the "reference" RNA, which is not of interest.

2.3.3 Randomization

The principle of randomization says that once any blocking structure to a design is established, treatments should be applied to experimental units in random fashion. If three littermates are to be divided among treatments A, B, and C, then the mice should be randomly allocated to each treatment. "Random" here does not mean the same thing as "arbitrary." Although tedious, it is useful to assign numbers to each mouse and use a random-number generator or draw numbers out of a hat to choose the mouse for each treatment.

While blocking protects against known or anticipated biases in the data, randomization protects against unknown or unanticipated biases. For the previous example, suppose one had an unrecognized tendency to pick-up the slowest mouse out of a litter. If one assigned mice to treatments A, B, and C in sequence, treatment A mice would tend to be assigned the slowest mice and treatment C would tend to be assigned to the quickest mice. If quick mice are also healthier, the experiment would obviously be biased.

Here is a more subtle, fictionalized example from the world of microarrays that shows that randomization is important even in observational studies. An experimenter is interested in a particular human mutation and recruits 20 carriers of the mutation. The mutation is rare and non-carriers are easier to find, and she is able to recruit 40 non-carriers to serve as controls. She is interested in whether the mutation is associated with any gene expression differences in humans. The investigator is reasonably confident that there are no other variables confounding the comparison between carriers and non-carriers. Using a single-color platform, the researcher uses one array to hybridize the mRNA for every individual. There is a practical limitation of a maximum of 20 hybridizations a day, so the experiment is carried out over three days.

The researcher applies a hierarchical clustering algorithm to explore the array data. The results appear as depicted in Figure 2.3(a). To the scientist's delight, the 60 samples appear to cluster into three primary groups: The 20 samples from the carriers of the mutation, and two groups of the remaining 40 non-carriers. The natural temptation is to conclude that gene expression data can discriminate carriers of the mutation from non-carriers, and that non-carriers can further be divided into two sub-types. However, with a healthy respect for scientific skepticism, the experimenter re-examines her data. Upon closer scrutiny, she sees that the three clusters correspond exactly to the three days of hybridizations, as in Figure 2.3(b).

In detail, the schedule for the hybridizations was:

- Day1: 20 carriers
- Day 2: 20 non-carriers
- Day 3: remaining 20 non-carriers

The fatal flaw in this investigation was the lack of randomization. The day of hybridization was ignored as a factor, but it turned out to be an important source of variation. Samples should have been hybridized in random order.

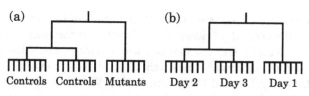

Fig. 2.3. Results of clustering samples for the example in Section 2.3.3. (a) Samples labeled by mutation status; (b) samples labeled by day of hybridization.

As is, the gene expression differences between carriers and non-carriers are hopelessly confounded with day-to-day differences in the hybridizations. There is no way to "rescue" the experiment – the confounding is complete and there is no way to separate the genetic differences of interest from the nuisance experimental artifacts.

Now that the day of hybridization is known to be an important factor, the researcher should probably "block" on the day of hybridization in future experimental plans. That is, for each group she should hybridize the same number of samples on each day.

2.4 Case Study

A plant geneticist is interested in the effects on gene expression in arabadopsis arising from infection by an agrobacterium. He plans a basic class comparison microarray study. From his initial collection of 20 plants, he randomly divides them into treatment and control groups of size 10. The treatment group is infected with the agrobacteria. The control group receives "mock" treatment, undergoing each step of infection except the introduction of the bacteria. This is to make sure that differences between the groups can properly be ascribed to infectious agent. One treated and control sample are produced every day, in random order. The RNA is extracted from each, and the treated and control RNA with same-day preparation are co-hybridized to a pair of microarrays employing dye-swap. That is, the design in Figure 2.2(a) is used, which is a very efficient design for comparing the two groups (Kerr, 2003a). This design will naturally handle any day-to-day differences in sample preparation (blocking) because day-to-day differences will cancel out in the treatment-control comparison due to the balance in the preparation schedule.

2.5 Conclusions

Replication, blocking, and randomization should all be considered in designing a microarray experiment. It usually works to consider them in the order presented here. First, make sure there is the right kind of replication to allow the desired inferences. Replication leads directly to the question of choosing a sample size. Sample size calculations are a tricky issue with microarrays and the subject of considerable research, beyond the scope of this article. See Simon et al. (2002); Lee and Whitmore (2002); Wei et al. (2004); and Tibshirani (2005). Second, for two-color platforms the arrangement of the samples onto the arrays must be decided. For many class comparison experiments the layouts in Figure 2.2 can be adapted. See Rosa et al. (2005), for other ideas. Lastly, consider all opportunities for randomization. For example, arrays can be randomly assigned to planned hybridizations and the order of hybridizations should also be randomized.

Although microarray studies are typically exploratory, one should still be able to clearly articulate a goal for the project. A well-defined goal will inform good choices in experimental design. A seriously flawed experimental design guarantees a study will be a failure, because it produces data that cannot answer the scientific question of interest. A sound experimental design does not guarantee a study will be a rousing success, but gives it a fighting chance.

References

Allison, D.B., Cui, X., Page, G.P., and Sabripour, M. (2006). Microarray data analysis: From disarray to consolidation and consensus. *Nat. Rev. Gen.*, 7:55–65.

Dobbin, K. and Simon, R. (2002). Comparison of microarray designs for class comparison and class discovery. *Bioinformatics*, 18(11):1438–1445.

Dudoit, S., Fridlyand, J., and Speed, T. P. (2002). Comparison of discrimination methods for the classification of tumors using gene expression data. *J. Am. Stat. Ass.*, 97:77–87.

Eisen, M.B., Spellman, P.T., Brown, P.O., and Botstein, D. (1998). Cluster analysis and display of genome-wide expression patterns. *Proc. Natl. Acad. Sc. USA*, 95(25):14863–14868.

Kerr, M.K. (2003a). Design considerations for efficient and effective microarray studies. *Biometrics*, 59:822–828.

Kerr, M.K. (2003b). Linear models for microarray data analysis: Hidden similarities and differences. *J. Comp. Biol.*, 10:891–901.

Kerr, M.K. and Churchill, G.A. (2001). Experimental design for gene expression microarrays. *Biostatistics*, 2:183–201.

Kerr, M.K., Martin, M., and Churchill, G.A. (2000). Analysis of variance for gene expression microarrays. *J. Comp. Biol.*, 7:819–837.

Lee, M.L. and Whitmore, G.A. (2002). Power and sample size for DNA microarray studies. *Statistics in Medicine*, 21(1):3543–70.

Potter, J.D. (2003). Epidemiology, cancer genetics and microarrays: making correct inferences, using appropriate designs. *TRENDS in Genetics*, 19(12):690–695.

Rosa, G.J.M., Steibel, J., and Tempelman, R.J. (2005). Reassessing design and analysis of two-colour microarray experiments using mixed effects models. *Comparative and Functional Genomics*, 6(1):123–131.

Simon, R., Radmacher, M.D., and Dobbin, K. (2002). Design of studies using DNA microarrays. *Gen. Epidem.*, 23:21–36.

Tibshirani, R. (2005). A simple method for assessing sample sizes in microarray experiments. http://www-stat.stanford.edu/ tibs/SAM/.

Wei, C., Li, J., and Bumgarner, R.E. (2004). Sample size for detecting differentially expressed genes in microarray experiments. *BMC Genomics*, 5:87.

Wolfinger, R.D., Gibson, G., Wolfinger, E.D., Bennett, L., Hamadeh, H., Bushel, P., Afshari, C., and Paules, R.S. (2001). Assessing gene significance from cDNA microarray expression data via mixed models. *J. Comp. Biol.*, 8:625–637.

Yang, Y.H. and Speed, T. (2002). Design issues for cDNA microarray experiments. *Nat. Rev.*, 3:579–588.

3

Pre-Processing DNA Microarray Data

Benjamin M. Bolstad

Department of Statistics, University of California, Berkeley, CA 94720-3860, USA.
bmb@bmbolstad.com

3.1 Introduction

Every microarray experiment produces images. Image analysis software reduces these images to raw intensity data. To be useful for a data analyst this raw intensity data need to be converted into gene expression measures. Pre-processing is used to describe these procedures. Note the terms "preprocessing", "low-level analysis", and "probe-level analysis" are synonymous and will be used interchangeably within this chapter. Some writers use the term "normalization" to encompass all the procedures discussed in this chapter, but here normalization will refer only to one stage of the process.

Unfortunately, many users of microarrays treat low-level analysis as a "black box", using whatever software is supplied by their system vendor, without much idea of what is really being done with their data. This chapter will highlight the importance of pre-processing, why a data analyst should know what is being done by the software and how it can improve subsequent data analysis.

Microarray experiments are usually conducted to answer one or more questions of biological interest, for instance topics such as: Determining gene function, discriminating between cases and controls or tumor sub-classes, studying the cell cycle and pathway analysis. Typically, low-level analysis methodologies do not attempt to answer these questions. Instead, the primary goal of a low-level analysis of a microarray data experiment is to provide better expression measures which can be used in higher-level analysis. Ideally, expression values should be both precise (low variance) and accurate (low bias). Another equally important aspect of low-level analysis is to be able to assess the quality of the microarray data.

Figure 3.1 shows the complete analysis process for a microarray experiment. After starting with a biological question, a sensible experiment is designed and carried out using microarrays. Images are produced as a result of the experiment and these are quantified to produce intensity values. The topic of this chapter, low-level analysis, is the next process. Consisting of both

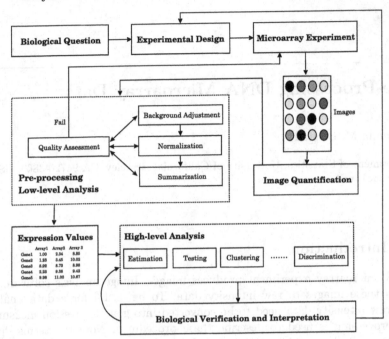

Fig. 3.1. Workflow for a typical microarray experiment.

steps to produce expression values and quality assessment it provides the data analyst with their first major task. If data from a particular array, or even worse the entire data set, is of low quality, then additional microarray experiments may need to be carried out. The expression values which result from the pre-processing analysis may then be used in higher-level analysis to try to directly answer the biological question of interest. This analysis may either lead to further high-level analysis or perhaps raise further biological questions that might warrant another microarray experiment.

Most microarray data is one of two basic flavors: *Single channel* or *two channel*, with the number of channels referring to the number of labeling colors used. On single channel arrays fragmented labeled RNA from a single source is hybridized to microarray, with the "rawest" level of data for each chip being a single image and the end result of the pre-processing analysis being an absolute measure of gene expression. Affymetrix GeneChip microarrays are a primary example of single channel microarrays. Two-channel microarrays have fragmented labeled cDNA from two sources with each source being labeled with a different color hybridized to microarray. These two colors are typically red and green, sometimes referred to using the dye names Cy5 and Cy3 (although other dyes are possible). For two-color microarray systems the "rawest" data consists of two images, one for each color channel. The end result of the pre-processing analysis is a measure of relative gene expression between the two samples directly compared on the slide. cDNA microarrays are a primary example of two-color microarrays, but long oligonucleotide arrays such as those commercially produced by Agilent and other companies also have two channels. This chapter uses the terminology *spotted array* and *two-color array* interchangeably.

While it might seem that these two basic platforms are quite different, both require the same basic pre-processing steps. In particular, *background correction, normalization, summarization* (this is applicable only in the case of Affymetrix GeneChips) and *quality assessment* are vital parts of low-level analysis for both single and two-channel microarray data.

This chapter introduces in some depth procedures and algorithms for pre-processing microarray data. Then two case studies, one using Affymetrix GeneChip data and the other using two-channel microarray data are used to illustrate these techniques in practice. The data used is available from the GEO data repository (Edgar et al., 2002; Barrett et al., 2005).

But before examining low-level analysis algorithms and their application to microarray data sets, it is important that the data analyst understands a little more about the underlying microarray technologies.

3.1.1 Affymetrix GeneChips

Affymetrix GeneChip arrays are a popular commercially produced high-density oligonucleotide array system produced using a photo-lithographic procedure. First, some sequence information of the target organism must be known. However, this does not present a particular difficulty because a number of organisms have now been completely sequenced and others are currently being sequenced. Given a known sequence, a number of 25-mer sequences complementary to the sequence for target genes are chosen. These sequences are known as *probes*. Typically, 11 to 20 probes interrogate a given gene. This collection of probes is called a *probe set* and there are currently anywhere between about 12 000 and 55 000 probe sets on an array. Affymetrix uses a number of procedures to select which 25-mer sequences should be used for each gene. In particular, potential probes are examined for specificity, potential for and predicted binding properties. Cross-hybridization occurs when a single-stranded DNA sequence binds to a probe sequence which is not completely complementary. To match the properties of the sample amplification procedure, probes are 3' biased. This means that probes are chosen closer to the 3' end of the sequence. However, the probes are typically spaced widely along the sequence. More details about probe selection are described in Mei et al. (2003). Sometimes there is more than one probe set that interrogate the same gene, but each uses a different part of the sequence.

On a GeneChip there are two types of probes. A probe that is exactly complementary to the sequence of interest is called a *Perfect Match* (PM). A probe that is complementary to the sequence of interest except at the central base, which for 25-mers is the 13th base, is known as the *Mismatch*

(MM). In theory, the MM probes can be used to quantify and remove non-specific hybridization. A PM and its corresponding MM probe are referred to as a *probe pair*. On modern chips the PM and its corresponding MM probe are contiguous on the microarray, but the probe pairs are distributed across the microarray. This helps alleviate potential problems from some types of spatial defects. Current microarrays have anywhere between 0.5 million and 2.5 million probes.

Once probes have been selected the chips are mass produced using a photo-lithographic procedure. Using a series of masks the 25-mer probes are built in parallel base by base.

Target mRNA source material is prepared in a series of transcribing, fragmenting, and labeling steps. It is then combined with the microarray in a process called *hybridization*, where using the complementary binding properties of DNA and RNA, sample material joins with the probes on the microarray.

After the washing and staining process the array is removed from the fluidics station and placed in a scanner. Laser light is shone onto the array and excites the fluorescent staining agent. At locations where more cRNA hybridized a brighter signal should be emitted. The amount of signal emitted is recorded as a value in 16 bits, and by examining the entire chip an image is produced. The Affymetrix software stores this image in the DAT file.

A small portion of the DAT file image is shown in Figure 3.2.

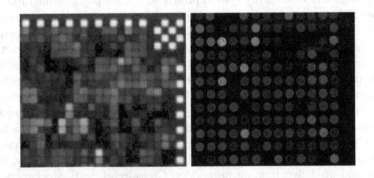

Fig. 3.2. Small sections of the raw image for an Affymetrix GeneChip as stored in a DAT file and the raw image for an Agilent two channel spotted array as stored in a TIFF file.

The checker board pattern and bright spots along the edges correspond to control oligo B2 probes. These are used to superimpose and align a grid upon the image. The grid is used to define the location of each probe cell. Each grid square contains all the pixels for a single probe. Once the gridding has taken place, the border pixels are ignored and the internal pixels of each grid square are used to compute a probe intensity. The probe intensity value for each probe cell is given by the 75th percentile of the intensities for these internal pixels. These probe intensity values are written into the CEL file. The low-level analysis we discuss in this chapter begins with data contained the CEL file.

3.1.2 Two-Color Microarrays

Spotted microarrays have thousands of individual DNA sequences or long oligos printed on a glass slide. This may be done using a robotic arrayer in the case of cDNA microarrays, inkjet technology in the case of Agilent microarrays or otherwise. Spots are typically arranged on the array in regularly spaced grids. The multiple grids on each array are typically each printed or spotted using a separate print-head. Depending on the slide design there may be control spots on the slide. Positive controls are spots which should be expressed in any sample hybridized to the array. Negative controls should not be expressed and are often based on sequences from another organism.

Two mRNA samples are transcribed to cDNA, labeled using different colored fluorescent dyes, mixed in equal proportions then and hybridized to the microarrays. After hybridization the microarray slide is scanned and fluorescence measurements made in each color channel producing an image. Unlike single channel arrays the hybridization is competitive with a particular spot being red, green or yellow corresponding to over-expressed or under-expressed in the red labeled sample or equal levels of expression in each sample. The ratio of the intensities in the red and green channels for each spot is intended to represent the relative abundance of the corresponding material in the two samples.

Image analysis for two-color arrays consists of three distinct stages. The first is addressing, or estimating the location of the center of each spot. Next a process called *segmentation* is used to classify pixels as either foreground (signal) or background. The final step is to extract spot information like the red and green channel foreground and background intensities and spot quality metrics. There are a number of different image analysis programs including GenePix, Spot, ScanAlyze, UCSF Spot and Imagene. A Web site listing additional microarray image analysis programs can be found in the list of tools and resources. This chapter is too brief to discuss the specific differences between these programs and instead the reader should refer to Yang et al. (2001, 2002a). The pre-processing methods discussed in this chapter begin with the output of these image programs.

3.2 Basic Concepts

Background correction is required for microarray data because typically there is some level of binding producing detectable signal even when that specific biological material is not in the original sample. Background correction methods typically attempt to perform one or more of the following: Correct for background noise and processing effects on the slide, adjust for cross-hybridization, which is the binding of non-specific DNA (i.e., non-complementary binding) to the probes on the array, or adjust expression estimates so that they fall on the proper scale (specifically, an even linear relationship between concentration of a given mRNA and its estimated expression is desirable).

Normalization is the process of removing unwanted non-biological variation that might exist within and between arrays in a microarray experiment. It has long been recognized that variability can exist between arrays, some of biological interest and other of non-biological interest. These two types of variation are classified as either interesting or obscuring by Hartemink et al. (2001). It is this obscuring variation that we seek to remove when normalizing arrays. Sources of obscuring variation can include scanner setting differences, the quantities of mRNA hybridized, dye labeling efficiencies and many other factors. Hartemink et al. (2001) discuss some of these possible sources in more detail.

Typically, Affymetrix GeneChip microarrays have hundreds of thousands of probes. These probes are grouped together into probe sets. Within a probe set each probe interrogates a different part of the sequence for a particular gene. Summarization is the process of combining the multiple probe intensities for each probe set to produce an expression value estimate.

Quality assessment is an important part of a low-level analysis involving both data inspection and decision making. Microarray data quality assessment can be carried out at several levels. For two-color arrays image analysis software often produce measures of individual spot quality and some users choose to integrate this into their analysis. A higher level is to consider quality at the array level, with the primary question being whether or not data from a particular microarray should be discarded and possibly repeated on an additional array.

3.2.1 Pre-Processing Affymetrix GeneChip Data

This section discusses the *robust multi-chip analysis* (RMA), methodology (Irizarry et al., 2003) and its extensions as the recommended procedures for probe-level analysis of GeneChip data. Some alternative methods are described later in the chapter. RMA uses only the PM probe intensities and ignores the MM probe intensities. This is because typically about 30% of MM intensities are higher, often significantly higher, than their corresponding PM intensity, and this is true across the entire range of probe-intensities. Simply subtracting the Mismatch intensity from the corresponding Perfect-match intensity would lead to negative expression values and an increased level of noise in expression values for low-expressing probe sets.

The background correction procedure for the RMA algorithm is based on a convolution model. Mathematically, a convolution model describes the distribution of the sum of two independent random variables. In particular, each observed PM probe intensity is assumed to be composed of a signal and noise component. By examining smoothed density histograms of intensity values from many GeneChips the suggested model consists of a normal distribution noise component and an exponential distribution signal. To avoid negative values the normal noise component is assumed to be truncated at 0. The background-corrected PM intensity is given by the expected value of the signal under the model given the observed PM intensity, a closed form equation for this expectation is given in Section 3.10. The parameters for each of the distributions are estimated on a chip-by-chip basis, so each chip gets an individual correction.

Next, RMA seeks to reduce non-biological variability by normalizing the background-corrected PM probe intensities across all the arrays in the data set. Because there are many hundreds of thousands of PM probe intensities on each array, and a data set may have many arrays, it is important that a normalization procedure is fast and scalable. The approach that RMA uses is the *quantile normalization* algorithm. The goal of quantile normalization, as discussed in Bolstad et al. (2003), is to give the same empirical distribution of intensities to each array. In other words, after quantile normalization the histogram of intensities on each array will be identical. The target distribution is found by averaging the quantiles for each of the arrays in the data set. A mathematical description of this transformation is given in Section 3.10. Quantile normalization is fast and scales well, with the algorithm consisting only of sorting, averaging and unsorting operations. In practice, quantile normalization performs very well at reducing unwanted variation. A thorough comparison of quantile normalization with other methods, and its effects on variability and bias, can be found in Bolstad et al. (2003).

As noted, the multiple PM probes on each Affymetrix GeneChip array targeting the same gene are known as a probe set. Gene expression summary values are found for each probe set by combining these background-corrected and normalized PM intensities. The RMA algorithm first $\log_2$ transforms the perfect match intensities. Then for each probe set a *multi-array probe-level model* (PLM) is fit using the *median polish algorithm*. An equation for the RMA PLM can be found in Section 3.10. The RMA model was suggested by examining probe intensity behavior within a probe set across arrays as shown in Figures 3.3 and 3.4. In these figures, lines have been used to join the intensities for specific probes across arrays. Notice that generally speaking the highest intensity probe is always the brightest and the lowest intensity probe the dimmest with the other probes falling in pretty much the same order in between. Because of this the RMA model includes an array effect and an effect for each probe and the transformed, background-corrected and normalized PM probe intensities are the response variables. The median polish algorithm is used to fit this model robustly. Robustness is needed because sometimes a particular probe intensity on a single array seems to behave quite differently than the corresponding probe on other arrays in the data set. The estimated array effect gives the estimated RMA expression value.

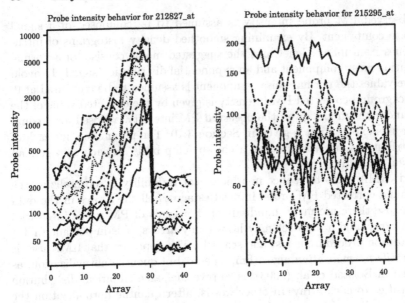

Fig. 3.3. Probe behavior for a spike-in probe set from an Affymetrix HG-U133A data set where by design there should be changes in expression level.

Fig. 3.4. Probe behavior for another probe set from an Affymetrix HG-U133A data set where there should be no differential expression.

The RMA framework can be extended to provide useful quality assessment criteria for when the computed gene expression measures on a particular array in a data set are of lower relative quality than the rest of the data set. The main changes are that an alternative model fitting procedure is used to fit the PLM. Specifically, the model is fit using robust regression via M-estimation (Huber, 1981) and estimates are generated using iteratively re-weighted least squares. This works by down-weighting an observation when it has a large residual from the fitted model and refitting the model again. This proceeds until convergence. As a by-product of the fitting procedure the final weights, residuals, standard error estimates for the expression values and the expression values themselves may be used to assess the quality of the data on each array in the data set. Typically, the weights or residuals are used to create chip pseudo-images to look for possible spatial artifacts and other differences between arrays. These tend to be better for visually detecting problems than images of the unprocessed probe-intensities. This is because the large differences in magnitude between the lowest and highest probe intensities are accounted for by the inclusion of the probe effect term in the model. Because there can be extreme differences in within probe set variability between probe sets the standard errors are converted into a measure called the *normalized unscaled standard error*, or NUSE for short. For each probe set the standard error estimates are normalized to have median 1 across arrays. An equation for doing this can be found in Section 3.10. An array of lesser quality is expected to have NUSE values that differ from others in the data set by being higher and possibly more variable. A second quantity, the *relative log expression* (RLE), is also used to assess the quality of the expression measures produced on a particular chip. This quantity is computed by comparing the expression value for a specific probe set on a particular array to the median expression value for that probe set across all the arrays in the data set. An equation for computing the RLE is given in Section 3.10. Arrays of lesser quality tend to have RLE values non-centered around zero or with greater spread. Typically, NUSE and RLE values are examined using boxplots, though some prefer to use numerical summaries such as the median and *interquartile range* (IQR) of these values for each array. The median provides a measure of center and the IQR provides a measure of spread. More details about using these quantities for quality assessment can be found in Brettschneider et al. (2006).

3.2.2 Pre-Processing Two-Color Microarray Data

Two-channel microarray data also require background correction, normalization and quality assessment. However, it has the additional complication that there are now two intensity channels to be dealt with.

Pre-processing begins with the output from the image analysis software. The export format of this data file varies depending on the image software used. But it typically includes measures of foreground and background intensity at each spot for each of the red and green channels along with a number of spot quality metrics.

Before pre-processing an initial quality analysis can be carried out. Spatial plots of the raw foreground and background intensities for each of the individual color channels and for spot statistics from the image analysis programs may show potential problems. Based on these it might also be useful to examine the raw TIFF files. Signal-to-noise (S2N) ratios, typically the log ratio of the foreground to background for each channel, should also be examined. Any control spots on the slide can also be used for this purpose

Background correction is the initial step in the pre-processing analysis. A first option is to simply subtract, in each channel, the background values from foreground values based on the image analysis output. However, this is typically problematic because most image analysis software make local estimates of background based on the pixels immediately surrounding the spot and these values tend be noisy. This leads to several problems, the first being that sometimes a spot will have a higher background value than foreground value, and so subtraction leads to a negative value, making log transforms impossible and leading to missing data. The second is that this background correction tends to inflate the noise in the expression values for low expressing genes. Instead, a better approach is to use a more smoothed estimate of background. The recommendations of Yang et al. (2002a) are that a morphological background such as those produced using Spot and recent versions of GenePix should be used instead. Others, such as Ritchie (2004), have explored the normal noise exponential signal convolution model and have found it to perform satisfactorily. All background corrections tend to increase the noise level, even if only slightly, of low expressed genes and because of this some users choose not to use any background correction.

Next, two-channel spotted array data need to be normalized. However, unlike single channel array data, normalization is carried out both within and between slides.

Within a slide, of particular interest are systematic differences due to intensity and location dependent dye biases. One way to compare red and green channel measurements for each spot on an array is an *MA-plot*. The M value on the vertical axis is the $\log_2$ ratio of the red and green channel intensities, in other words the $\log_2$ fold-change between the two samples. The A value on the horizontal axis is the average of the $\log_2$ red and green channel intensities. Often a loess smoother is added to show general trends in the data. Figure 3.5 shows a typical MA-plot for a two-channel microarray. A smooth loess curve is typically added to show the general trend of the data. Loess is a method of local regression due to Cleveland and Devlin (1988) that is used for estimating a smooth non-parameteric curve.

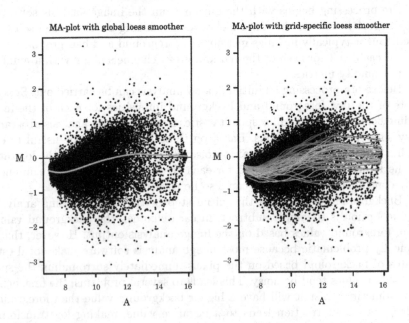

Fig. 3.5. MA-plots for unnormalized two channel data. The MA-plot on the left has a single loess smoother showing global intensity dye biases. The MA-plot on the right has the loess smoothers for each of the grids on this particular microarray

An MA-plot is more useful for examining the relationship between the two channels than other alternatives. For instance, a plot of the red against green intensities will be misleading with microarray data because there is typically very few bright spots on an microarray and many dim spots meaning many points in the lower left hand region of the plot. Additionally, any correlation calculated will be high and driven mostly by the few extreme points. A better alternative would be to log transform both the red and green intensities and then plot them against each other. But, this plot still lives much of the plotting region unused and visually the linear relationship between the two channels might still appear strong. An MA-plot is essentially a rotation of this log-transformed plot so that the 45 degree line is now the horizontal axis.

Assuming that only a few genes are changing in expression or that at least about equal numbers of genes should increase and decrease in expression between the two samples hybridized to the array then ideally the point cloud on the MA-plot will be centered around 0. The loess curve on the MA-plot can be used to give an intensity dependent correction. After fitting the loess curve to the MA-plot each M value is adjusted vertically up or down by the amount required to adjust the loess curve at the corresponding A value so that it falls along the $M = 0$ horizontal axis. This normalization removes global intensity dependent dye biases and is called *loess normalization method*.

As demonstrated in Figure 3.5, it is also possible that there are spatial differences in this intensity dependent dye bias and other effects caused by different printing-tips. Multiple loess curves, one for each grid, are fitted. The M values in each grid are adjusted in the same manner as before using the loess curve specific to that grid. This is known as *print-tip loess normalization*.

The loess normalization methods adjust for location, but it is also possible that there are regional differences in variability. In other words, the M values have differing variability depending upon which grid they lie in. A scale normalization is proposed by Yang et al. (2002a) which standardizes the median absolute deviation of M values for each grid.

Most two-color microarray experiments consist of multiple arrays, and because there can be many sources of technical variability, *between-slide normalization* is also required. Between-slide normalization should be carried out after *within-slide normalization* is completed. The scale normalization proposed for within-array normalization could be used for normalizing the M values between arrays. Another option is to normalize the individual red and green channel intensity data across arrays. The quantile normalization procedure used for the Affymetrix GeneChip data can be used for this purpose. In particular, the $\log_2$ scale red and green channel intensities are normalized across all arrays.

Quality assessment should be integrated at all steps of the pre-processing process. Usually, this is done visually. Boxplots of M values before and after normalization show removal of some differences. Spatial plots of quantities including M values, show possible spatial effects and other artifacts such as scratches on the slide. MA-plots are also useful for examining the degree of normalization required. If after applying the different pre-processing procedures the quality assessment methods still indicate problems then data from that microarray may need to be discarded.

3.3 Advantages and Disadvantages

3.3.1 Affymetrix GeneChip Data

The RMA methodology and its PLM based extensions have the following advantages and disadvantages.

3.3.1.1 Advantages

- Fast and scalable to large data sets.
- Widely accepted and used by many researchers.
- RMA is available in a number of software packages including Bioconductor, GeneSpring, ArrayAssist, S+ArrayAnalyzer, GeneSifter.
- Has little noise at low intensities.
- The quality assessment measurements are directly related to the quality of the produced expression values.
- Improves ability to correctly identify differentially expressed genes and is highly reproducible.
- Robust against outliers when computing expression measure.

3.3.1.2 Disadvantages

- RMA has a tendency to attenuate fold-change estimates for low-expressing probe sets.

3.3.2 Two-Color Microarrays

The techniques for pre-processing spotted microarray data have the following advantages and disadvantages.

3.3.2.1 Advantages

- Loess-based normalization deals effectively with non-linearities in the data.
- Combining quality assessment with normalization allows the analyst to see which problems are being accounted for by normalization and those that are because of bad quality data.
- Can deal with some spatial effects.
- Attempts to deal with both within-slide and between-slide technical variability.

3.3.2.2 Disadvantages

- Not standard across software packages. For instance, many software programs do not produce MA-plots.

3.4 Caveats and Pitfalls

To some degree most standard normalization methods, on both Affymetrix and two-color microarrays, depend on at least one of the following assumptions:

1. The number of genes changing between conditions is small relative to the number of genes being measured on the microarray.
2. An approximately equivalent number of genes are increasing in expression value as are going down in expression value between conditions.

If either of these assumptions are violated then there is the possibility that small changes in expression might be made undetectable, and larger changes made smaller. In the case of the print-tip loess normalization, not only should these assumptions be true across the entire array, but also within each grid. However, in the vast majority of situations these assumptions hold, if only weakly.

In situations where these assumptions do not hold, one option would be to normalize within arrays of the same condition with a stronger method, say quantile normalization for instance, and between conditions with a weaker method, such as scaling so that the means or medians are equal. It is usually preferable to use some level of normalization, than to not normalize at all.

3.5 Alternatives

3.5.1 Affymetrix GeneChip Data

RMA is not the only expression measure possible. A popular modification of the algorithm is known as GCRMA (Wu et al., 2004). It uses a different background correction algorithm that incorporates probe sequence information, but uses the same quantile normalization and median polish summarization procedure. This has been observed to improve the problems RMA has with attenuating expression values, with only a small increase in variability.

The dChip MBEI (Li and Wong, 2001a,b) also uses a multi-array model; however, it differs from the RMA model in that the model is multiplicative with additive errors fitted on the natural scale and a different non-linear normalization algorithm is used. This tool also provides a method for visually assessing quality by identifying outlier probes.

Affymetrix provides the MAS 5.0 expression measure (Affymetrix, 2001) and more recently an algorithm called PLIER may also be used to generate gene expression values. MAS 5.0 values are typically noisy in the low-intensity range and use a simple linear scaling normalization. There are a number of standard quality assessment quantities produced by the Affymetrix software including the percent present, scale factor, average background and 3'/5' ratios for control probe sets.

There is some contention about which gene expression measure is the best to use and there are many alternatives beyond those discussed in this chapter. However, it is possible to measure and quantify the effect that these algorithms have on data where there is known differential expression. Spike-in data sets are used for this purpose. The affycomp benchmarking tool of Irizarry et al. (2006) provides a basis upon which to assess the performance of different methods in terms of bias, variance and ability to detect differential expression.

3.5.2 Two-Color Microarrays

Some users choose to use control spots on the microarray for normalization on the array. This can be potentially problematic if the control spots are not typical for other spots on the array. For instance they could all be on average brighter than the majority of spots on the array.

Depending on the slide design it might be possible to combine the loess-based methods which use all spots with information from the control spots. Yang et al. (2002b) suggest using a Microarray Sample Pool (MSP), where control spots are created using cDNA for all target genes on the array and then spotted in a dilution series. These are then combined with the loess normalization method so that dim spots are primarily normalized using the loess normalization and higher intensities use progressively greater information from the MSP spots.

Many microarray analysis programs implement basic scaling normalization algorithms, where the mean or median value in each color channel are made equal by multiplying by a constant. In general these should be avoided in preference for methods which can deal with non-linearities.

Variance stabilization normalization (VSN) provides another alternative to the loess based methods. In particular this normalization method seeks to make the variance of the data independent from the mean. This is done using a generalized log transformation. For further details, see Huber et al. (2002).

3.6 Case Study

3.6.1 Pre-Processing an Affymetrix GeneChip Data Set

In this section a typical pre-processing analysis for an Affymetrix GeneChip data set is demonstrated. For illustrative purposes data from 25 HG-U133A microarrays, a subset of a larger data set, is used. This data was down-loaded from the NCBI Gene Expression Omnibus (GEO, http://www.ncbi.nlm.nih.gov/geo/) and is accessible through GEO Series accession number GSE2603. The purpose of the study that generated this data was to discover a set of genes which mark and mediate breast cancer metastasis to the lungs. Further details about this can be found in Minn et al. (2005).

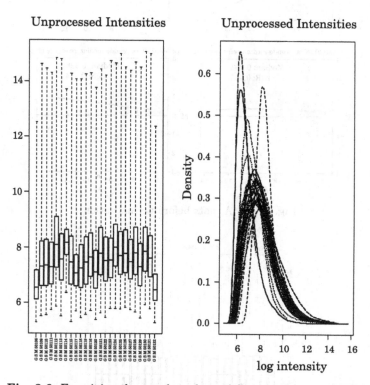

Fig. 3.6. Examining the raw data shows differences between arrays.

Initially, the raw intensities are examined by array to look for differences. Figure 3.6 shows boxplots and density plots of raw $\log_2$ PM intensities for all 25 arrays. It is immediately apparent that there are differences in intensity level between the arrays. Two arrays, GSM50108 and GSM50132, are dimmer than the others, another (GSM50112) seems to be brighter.

Rather than look at MA-plots for every pair of arrays, a synthetic reference chip is created by taking probe-wise medians and then each array is compared to this reference. Figure 3.7 shows MA-plots for four of the arrays with each showing significant differences.

Having established that there are differences in the unprocessed probe-intensities, they are pre-processed and summarized to RMA gene expression

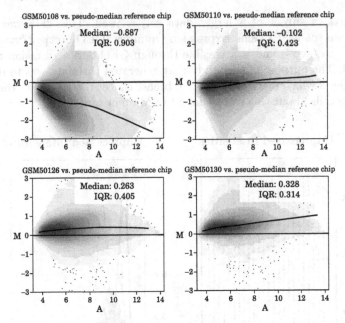

Fig. 3.7. MA plots before normalization.

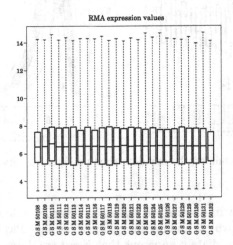

Fig. 3.8. Boxplot of RMA expression values by array show fewer differences than raw data.

values. Figure 3.8 shows boxplots of the RMA expression measures for each array with the differences now minimal compared to the raw unprocessed data. Figure 3.9 shows the MA-plots for the four arrays considered earlier. Two arrays, GSM50126 and GSM50130, show significant improvement after nor-

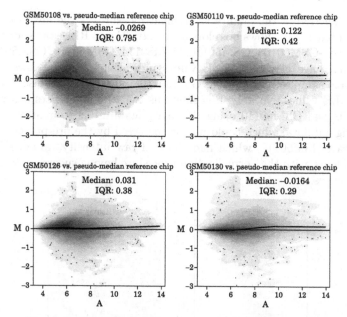

Fig. 3.9. MA plots based on computed expression values.

malization. However, GSM50108 still has a divergent MA-plot with elevated IQR, and the median M value for GSM50110 is away from 0, the ideal.

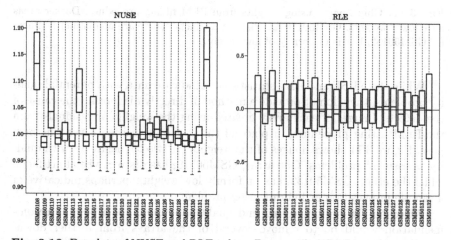

Fig. 3.10. Boxplots of NUSE and RLE values. Boxes with different centers or larger spreads indicate quality problems.

An inspection of the quality assessment measures is the next step in the analysis. Figure 3.10 shows boxplots of the NUSE and RLE statistics for each array. Arrays GSM50108, GSM50110, GSM50114, GSM50116, GSM50120 and GSM50132, have elevated NUSE values indicating potential lower quality data. Three samples, GSM50108, GSM50110 and GSM50132 also had problematic RLE values. Notably, this shows us that the differences observed for GSM50108 and GSM50110 using the MA-plots are due to quality problems in the original data, which could not be corrected by the pre-processing. Another array, GSM50112, which had elevated raw intensity values, shows no such problems, with the normalization procedure having successfully removed the differences.

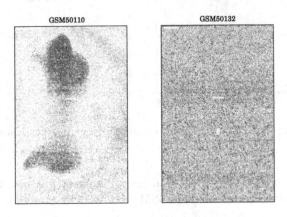

Fig. 3.11. Chip images using weights from PLM fitting procedure. Darker areas have lower weights. GSM50110 has distinct artifacts. There are no distinct artifacts for GSM50132, but is uniformly of low weight.

While the NUSE and RLE statistics can indicate data with potential quality problems, they do not provide a method of directly diagnosing the cause of the defects. Images of each array created using the weights or residuals from the PLM fitting procedure, in place of the probe intensities, serve this purpose. Figure 3.11 shows images of the weights for two of the arrays with quality assessment problems. One array, GSM50110, has distinct visual artifacts, while the other, GSM50132, has uniformly low weights, perhaps indicative of a sample preparation or hybridization problem.

Based on the pre-processing and quality assessment analysis the recommendation would be to remove several of the arrays from further downstream analysis. The evidence is particularly strong for removing GSM50108, GSM50110 and GSM50132. However, data from the arrays that had elevated NUSE values, but more normal RLE values, are also suspect.

3.6.2 Pre-Processing a Two-Channel Microarray Data Set

This section considers a typical pre-processing analysis for a two-color array data set. The data set used consists of 15 microarray slides, a subset of a much larger study. This data was retrieved from the NCBI Gene Expression Omnibus (GEO, http://www.ncbi.nlm.nih.gov/geo/) and is accessible through GEO Series accession number GSE1438. The specific slide design used, UCSF 10Mm Mouse v.2 Oligo Array (GEO GPL1089), has 4 columns and 12 rows of grids with a total of 18 240 spots on each array. The purpose of the study that generated this data was to identify how effects of IL-13 on airway epithelial cells contribute to gene expression changes in Murine asthma models. Lung gene expression was analyzed by hybridizing Cy5-labeled cDNA from mouse lungs (five mice per group, each hybridized separately) along with Cy3-labeled reference lung cDNA pooled from wild-type mice. The study also considered RNA from tracheal perfusate samples. Further details can be found in Kuperman et al. (2005). The pre-processing analysis in this section considers only data from the lung samples. The data was image analyzed using GenePix.

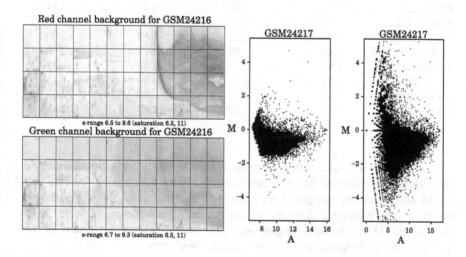

Fig. 3.12. Red and green channel background intensities show a visible artifact for GSM24216.

Fig. 3.13. MA-plots before and after local background correction.

Examining spatial plots (images) of the raw single channel background intensities show any spatial artifacts on the slide. Figure 3.12 shows an artifact in both channels for the slide GSM24216. Artifacts or high uniform background signals can indicate potential problems.

In this section, the pre-processing analysis is carried out without background correction. The reason for this is demonstrated in Figure 3.13 that shows how the noise is greatly inflated by carrying out local background correction. However, the background values can still be used to compute signal-to-noise values for each channel. Table 3.1 shows the mean and variance of the signal-to-noise ratios for all 15 slides in the data set. Several arrays, GSM24219, GSM24231 and GSM24232, have low signal-to-noise ratios in one of the channels. Several others such as GSM24220 and GSM24232 have large differences in the signal-to-noise ratio between the two channels.

Table 3.1. Signal-to-noise values for the red and green channels. Higher mean values are better. A mean value above 1.0 indicates that the signal is higher than the background.

Slide	mean $S2N_R$	var $S2N_R$	mean $S2N_G$	var $S2N_G$
GSM24215	2.61	3.33	2.42	3.23
GSM24216	1.35	2.28	1.43	2.30
GSM24217	1.23	1.77	1.67	2.31
GSM24218	1.58	2.15	1.78	2.41
GSM24219	1.27	2.24	0.94	1.13
GSM24220	2.50	3.83	1.79	2.35
GSM24221	1.89	3.00	2.21	3.54
GSM24224	2.39	2.90	2.36	3.03
GSM24225	2.21	3.21	1.88	2.00
GSM24229	2.06	2.82	1.72	2.06
GSM24230	1.95	3.03	1.29	1.51
GSM24231	1.72	2.70	0.99	1.11
GSM24232	1.03	1.56	1.48	2.05
GSM24233	1.69	2.89	1.49	2.23
GSM24234	1.73	2.20	2.05	2.67

The sample GSM24216 is used to show how the data for a single slide can be examined before and after normalization. Figure 3.14 shows the ranks of the M and A values computed for each spot on this array and then plotted spatially. Non-biological differences show up in uneven spread of ranks. Box-plots of the M values by grid position, as shown in Figure 3.15, also shows that the raw fold-change values have spatial dependence.

To correct for these technical differences, print-tip loess normalization is used. Figure 3.16 shows the individual loess curves, one for each of the 48 grids, used to normalize the data on the slide GSM24216. Differences between the loess lines are apparent. This demonstrates the need to use one for each grid.

Figure 3.17 shows the M values after print-tip loess normalization, both as a spatial plot of ranked M values and boxplots of the M values by grid. Many of the technical differences observed in the raw data due to spatial differences and dye-biases have been reduced.

The print-tip loess normalization method is designed to remove non-biological differences within slide. However, it also serves to reduce some of

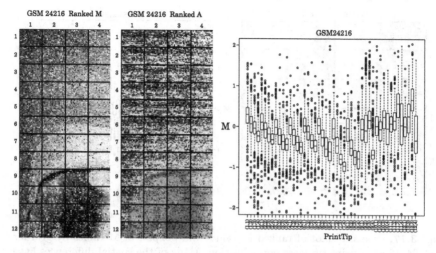

GSM 24216 Ranked M   GSM 24216 Ranked A   GSM24216

Fig. 3.14. Spatial plots of ranked M   **Fig. 3.15.** Boxplots of M values by grid
and ranked A values for GSM24216.   for GSM24216.
Artifacts are visible.

the variability across arrays. Figure 3.18 shows boxplots of the M values across
arrays *before*, and Figure 3.19 *after* normalizing within each slide using the

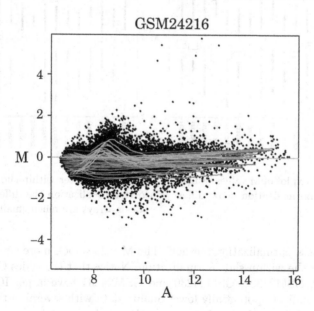

Fig. 3.16. Normalization using print-tip loess. Separate loess curves are fit for each
grid.

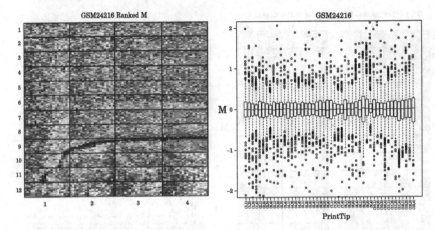

Fig. 3.17. A spatial plot of ranked M values and boxplot of M values by grid for GSM24216 after print-tip loess normalization. Many of the spatial differences have been removed although one artifact remains.

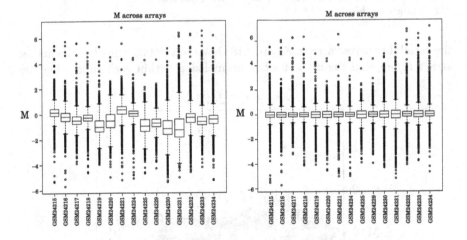

Fig. 3.18. Boxplot of M values by slide before any normalization show differences.

Fig. 3.19. After within-slide print-tip loess normalization the differences between arrays are much smaller.

print-tip loess normalization method. The M values look more similar across arrays after the within-slide normalization. Notice that the slides GSM24219, GSM24220, GSM24225, GSM24230 and GSM24231 have larger IQR values. These may indicate potentially lower quality data with several of these corresponding to slides with signal-to-noise ratio problems.

Although we saw that the M values seemed to be much less variable across arrays after within-slide normalization, there are still differences in the data

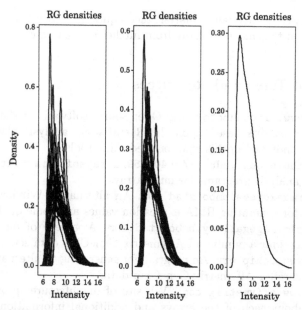

Fig. 3.20. Density plots of (left) red and green channel intensities before normalization, (middle) after within-slide print-tip loess normalization and (right) after between-slide single channel quantile normalization

that can be removed across arrays. Figure 3.20 shows density plots of the red and green channel intensities. The first plot shows the raw data with differences visible both between arrays and between channels. The second plot shows the situation after within-slide normalization. The differences between the red and green channels on each array have been reduced, but there are still some differences between arrays. The third plot shows the result after between-slide single channel quantile normalization applied following the within-slide normalization. In this case, the differences have been completely removed.

3.7 Lessons Learned

The two case studies both highlight how the pre-processing methodologies discussed in this chapter can be applied to real microarray data to remove technical biases and variability. Additionally, the case studies demonstrate that data set examination is an essential part of the process. Data analysts should use the plotting tools discussed in this chapter to visualize their data and treat it as equally important to just applying algorithms.

As we saw, in both the Affymetrix GeneChip and two-color case studies, pre-processing can not and should not be expected to correct for all possible problems. Instead, the quality analysis techniques described here can be used to decide when to remove an array from further downstream analysis.

3.8 List of Tools and Resources

- `http://www.bioconductor.org`: Open source software for the analysis of genomic data sets based upon the R statistical analysis platform. The analysis conducted in this chapter used these tools.
- `http://ihome.cuhk.edu.hk/~b400559/arraysoft_image.html`: Listing of image analysis programs for microarrays.
- `http://rmaexpress.bmbolstad.com`: An alternative Windows GUI application for generating RMA expression values and some quality images.
- `http://plmimagegallery.bmbolstad.com`: A gallery of quality assessment images for a number of Affymetrix GeneChip data sets.
- `http://www.dchip.org`: An program for generating an alternative expression measure for Affymetrix GeneChip data.
- `http://www.affymetrix.com`: Web site of manufacturer provides information about each of the arrays and additional information about the technology.
- `http://affycomp.biostat.jhsph.edu`: Benchmarking tool for comparing the performance of alternative expression measures for GeneChip data.
- `http://bmbolstad.com/FDMGP`: Supplemental material for this chapter including complete data analysis code and additional plots.

3.9 Conclusions

Pre-processing is a very important step in the examination of a microarray data set. The low-level analysis techniques discussed in this chapter help remove differences due to technical, rather than biological, differences. Gene expression estimates generated using sensible pre-processing will be more useful for higher-level analysis. While pre-processing can improve the quality of some data, the quality assessment techniques discussed in this chapter provide methods for deciding when to remove arrays from further analysis.

3.10 Mathematical Details

3.10.1 RMA Background Correction Equation

Assume that the observed PM intensity for any probe on an array consists of a signal and noise component. Specifically, the observed PM intensity Y is the sum of signal S and noise N, i.e., $Y = S + N$. Assume that S follows an exponential distribution with parameter α, and N is distributed normally (truncated at 0 to avoid negatives) with mean μ and variance σ^2. Then the background-corrected value is

$$E(s|Y = y) = a + b\frac{\phi\left(\frac{a}{b}\right) - \phi\left(\frac{y-a}{b}\right)}{\Phi\left(\frac{a}{b}\right) + \Phi\left(\frac{y-a}{b}\right) - 1} \quad, \tag{3.1}$$

where $a = y - \mu - \sigma^2\alpha$ and $b = \sigma$. Note that ϕ and Φ are the standard normal distribution density and distribution functions, respectively. In practice, the second term on the numerator is essentially 0 and the second term on the denominator is essentially 1, allowing the formula to be simplified.

3.10.2 Quantile Normalization

The goal of the quantile normalization algorithm is to give the same distribution of intensities to each array. An intensity is transformed in quantile normalization in the following manner:

$$x_{ij}^* = F^{-1}\left(G_j\left(x_{ij}\right)\right) \quad, \tag{3.2}$$

where x_{ij} is measurement i on array j, G_j is the distribution function for array j, and F^{-1} is the inverse of the distribution function to be normalized to. The normalized intensity is given by x_{ij}^*. In practice, G_j is estimated using the empirical distribution function and F is the average distribution across all arrays in the data set. An implementation of the algorithm is described in Bolstad et al. (2003).

3.10.3 RMA Model

The RMA expression measure is based upon fitting the following model

$$y_{kij} = \beta_{kj} + \alpha_{ki} + \epsilon_{kij} \tag{3.3}$$

on a probe set by probe set basis. The indices k, j and i refer to probe set, array and probe, respectively. The response terms y_{kij} are $\log_2$ transformed background-corrected and normalized PM intensities. The parameters β_{kj} and α_{ki} represent the chip effect and probe effect, respectively. The error term is ϵ_{kij}. To make the model identifiable it is fit with the constraint $\sum_{i=1}^{I_k} \alpha_{ki} = 0$. The estimates $\hat{\beta}_{kj}$ provide the $\log_2$ scale RMA expression values.

3.10.4 Quality Assessment Statistics

The NUSE quality assessment statistic is calculated for each probe set on each array using:

$$\text{NUSE}\left(\hat{\beta}_{kj}\right) = \frac{\text{SE}\left(\hat{\beta}_{kj}\right)}{\text{med}_j\text{SE}\left(\hat{\beta}_{kj}\right)} \tag{3.4}$$

The RLE quality assessment statistic is calculated for each probe set on each array using:

$$\text{RLE}\left(\hat{\beta}_{kj}\right) = \hat{\beta}_{kj} - \text{med}_j\beta_{kj} \tag{3.5}$$

3.10.5 Computation of M and A Values for Two-Channel Microarray Data

Suppose that R_i and G_i are respectively red and green channel intensities for spot i on a specific array. Then the M and A values for each spot are given by:

$$M_i = \log_2\left(\frac{R_i}{G_i}\right) = \log_2 R_i - \log_2 G_i \tag{3.6}$$

$$A_i = \frac{1}{2}\log_2\left(R_i \cdot G_i\right) = \frac{1}{2}\left(\log_2\left(R_i\right) + \log_2\left(G_i\right)\right) \tag{3.7}$$

3.10.6 Print-Tip Loess Normalization

If M_i represent the unnormalized M values on a slide, A_i the corresponding A values and there are $k = 1, \ldots K$ grids of spots on the array. Then the normalized log-ratios are given by

$$M_i^* = M_i - c_k\left(A_i\right) = \log_2\left(\frac{R_i}{G_i}\right) - c_k\left(A_i\right), \tag{3.8}$$

where $c_k\left(A\right)$ is the loess curve estimated using the M and A values corresponding to spots in grid k.

References

Affymetrix (2001). *Affymetrix Microarray Suite Users Guide*. Affymetrix, Santa Clara, CA, version 5.0 edition.

Barrett, T., Suzek, T.O., Troup, D.B., Wilhite, S.E., Ngau, W.C., Ledoux, P., Rudnev, D., Lash, A.E., Fujibuchi, W., and Edgar, R. (2005). NCBI GEO: Mining millions of expression profiles-database and tools. *Nucleic Acids Res.*, 33(Database issue):562–566.

Bolstad, B.M., Irizarry, R.A., Astrand, M., and Speed, T.P. (2003). A comparison of normalization methods for high density oligonucleotide array data based on variance and bias. *Bioinformatics*, 19(2):185–193.

Brettschneider, J., Collin, F., Bolstad, B.M., and Speed, T.P. (2006). Quality assessment for short oligonucleotide microarray data. Technical report, Queens University, Kingston, ON, Canada.

Cleveland, W.S. and Devlin, S.J. (1988). Locally-weighted regression: An approach to regression analysis by local fitting. *J. Am. Stat. Assoc.*, 83:596–610.

Edgar, R., Domrachev, M., and A.E., Lash (2002). Gene Expression Omnibus: NCBI gene expression and hybridization array data repository. *Nucleic Acids Res.*, 30(1):207–210.

Hartemink, A.J, Gifford, D.K., Jaakkola, T., and Young, R.A. (2001). Maximum likelihood estimation of optimal scaling factors for expression array normalization. In *SPIE BIOS 2001*.

Huber, P.J. (1981). *Robust Statistics*. John Wiley & Sons, Inc, New York, New York.

Huber, W., von Heydebreck, A., Sultmann, H., Poustka, A., and Vingron, M. (2002). Variance stabilization applied to microarray data calibration and to the quantification of differential expression. *Bioinformatics*, 18 Suppl. 1:S96–S104.

Irizarry, R.A., Bolstad, B.M., Collin, F., Cope, L.M., Hobbs, B., and Speed, T.P. (2003). Summaries of Affymetrix GeneChip probe level data. *Nucleic Acids Res.*, 31(4):e15.

Irizarry, R.A., Z., Wu, and Jaffee, H.A. (2006). Comparison of Affymetrix GeneChip expression measures. *Bioinformatics*, 22(7):789–794.

Kuperman, D.A., Lewis, C.C., Woodruff, P.G., Rodriguez, M.W., Yang, Y.H., Dolganov, G.M., Fahy, J.V., and Erle, D.J. (2005). Dissecting asthma using focused transgenic modeling and functional genomics. *J. Allergy Clin. Immunol.*, 116(2):305–311.

Li, C. and Wong, W.H. (2001a). Model-based analysis of oligonucleotide arrays: Expression index computation and outlier detection. *Proc. Natl. Acad. Sci. USA*, 98(1):31–36.

Li, C. and Wong, W.H. (2001b). Model-based analysis of oligonucleotide arrays: Model validation, design issues and standard error application. *Genome Biol.*, 2(8):RESEARCH0032.

Mei, R., Hubbell, E., Bekiranov, S., Mittmann, M., Christians, F.C., Shen, M.M., Lu, G., Fang, J., Liu, W.M., Ryder, T., Kaplan, P., Kulp, D., and Webster, T.A. (2003). Probe selection for high-density oligonucleotide arrays. *Proc. Natl. Acad. Sci. USA*, 100(20):11237–11242.

Minn, A.J., Gupta, G.P., Siegel, P.M., Bos, P.D., Shu, W., Giri, D.D., Viale, A., O., A.B., Gerald, W.L., and Massague, J. (2005). Genes that mediate breast cancer metastasis to lung. *Nature*, 436(7050):518–524.

Ritchie, M.E. (2004). *Quantitative quality control and background correction for two-colour microarray data*. PhD thesis, The Walter and Eliza Hall Institute of Medical Research, Melbourne, Australia.

Wu, Z., Irizarry, R.A., Gentleman, R., Martinez-Murillo, F., and Spencer, F. (2004). A model-based background adjustment for oligonucleotide expression arrays. *J. Am. Stat. Ass.*, 99(468):909–917.

Yang, Y.H., Buckley, M.J., Dudoit, S., and Speed, T.P. (2002a). Comparison of methods for image analysis of cDNA microarray data. *J. Comp. Graph. Stat.*, 11:108–136.

Yang, Y.H., Buckley, M.J., and Speed, T.P. (2001). Analysis of cDNA microarray images. *Brief. Bioinform.*, 2(4):341–349.

Yang, Y.H., Dudoit, S., Luu, P., Lin, D.M., Peng, V., Ngai, J., and Speed, T.P. (2002b). Normalization for cDNA microarray data: a robust composite method addressing single and multiple slide systematic variation. *Nucleic Acids Res.*, 30(4):e15.

4

Pre-Processing Mass Spectrometry Data

Kevin R. Coombes, Keith A. Baggerly, and Jeffrey S. Morris

Department of Biostatistics and Applied Mathematics, University of Texas M.D. Anderson Cancer Center, Houston, TX 77030, USA.
krc@odin.mdacc.tmc.edu, kabagg@wotan.mdacc.tmc.edu, jeffmo@wotan.mdacc.tmc.edu

4.1 Introduction

Mass spectrometry is being applied to discover disease-related proteomic patterns in complex mixtures of proteins derived from tissue samples or from easily obtained biological fluids such as serum, urine, or nipple aspirate fluid (Paweletz et al., 2001; Wellmann et al., 2002; Petricoin et al., 2002; Adam et al., 2002, 2003; Zhukov et al., 2003; Schaub et al., 2004). Potentially, we can use these proteomic patterns for early diagnosis, to predict prognosis, to monitor disease progression or response to treatment, or even to identify which patients are most likely to benefit from particular treatments.

The mass spectrometry instruments most commonly used to address these clinical and biological problems use a matrix-assisted laser desorption and ionization (MALDI) ion source and a time-of-flight (TOF) detection system. Briefly, to run an experiment on a MALDI-TOF instrument, the biological sample is first mixed with an energy absorbing matrix (EAM) such as sinapinic acid or α-cyano-4-hydroxycinnamic acid. This mixture is crystallized onto a metal plate. (The commonly used method of surface-enhanced laser desorption and ionization (SELDI) is a variant of MALDI that incorporates additional chemistry on the surface of the metal plate to bind specific classes of proteins (Merchant and Weinberger, 2000; Tang et al., 2004).) The plate is inserted into a vacuum chamber, and the matrix crystals are struck with pulses from a nitrogen laser. The matrix molecules absorb energy from the laser, transfer it to the proteins causing them to desorb and ionize, and produce a plume of ions in the gas phase. This process takes place in the presence of an electric field, which accelerates the ions into a flight tube where they drift until they strike a detector that records the time of flight (Figure 4.1).

In theory, the spectral data produced by a single laser shot in a mass spectrometer consists of a vector of counts. Each count represents the number of ions hitting the detector during a small, fixed interval of time. We refer to this interval of time as the *time resolution* of the instrument; the time resolution is typically on the order of 1–4 nanoseconds. A complete spectrum is

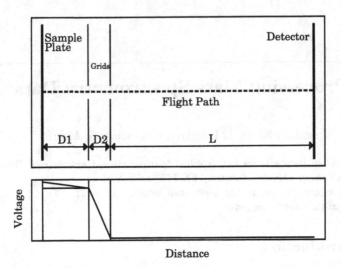

Fig. 4.1. (Top) Simplified schematic of a MALDI-TOF instrument with time-lag focusing. Samples are inserted on a metal plate into a vacuum chamber where they are ionized by a laser. Electric fields between the sample plate and two charged grids accelerate the ions into a drift tube, where they continue until they strike a detector. **(Bottom)** Voltage potentials along the instrument. The sample plate and grid start at the same potential, but the potential is raised after a brief delay.

acquired within tens of milliseconds, so a typical spectrum is a vector containing between 10 000 and 100 000 entries. In practice, most mass spectrometers produce spectra by averaging the counts over many (often a few hundred) individual laser shots. Thus, the raw data produced by running a sample through a mass spectrometer can best be thought of as a time series (see Chapter 11) vector containing tens of thousands of real numbers. Unless an entry in the vector is known to represent an actual count of the number of ions, it is usually just called an *intensity* and is assumed to be measured in continuous arbitrary units. Peaks in a plot of the intensity as a function of time represent the proteins or peptides that are present in the sample (Figure 4.2, top).

It is important to realize that the natural scale on which to view a mass spectrum is the time axis along which the data was originally collected. Applications of mass spectrometry are, however, based on the mass of the particles. Ions of different mass are separated in the flight tube. In general, lighter ions fly faster and thus reach the detector before heavier ions. More precisely, the velocity achieved by an ion is proportional to its *mass-to-charge ratio* (m/z). A quadratic transformation is used to compute m/z from the observed flight time. The coefficients of this quadratic transformation must be determined experimentally. Researchers prepare a sample containing a small number (typically between 3 and 7) of molecules of known masses and use it to generate a spectrum. They then determine the times at which the peaks corresponding to the known masses occur in that spectrum, and use least squares and this set of (*time, mass*) pairs to determine the coefficients of the quadratic transformation. The process of mapping the observed time of flight to the m/z values is called *calibration*.

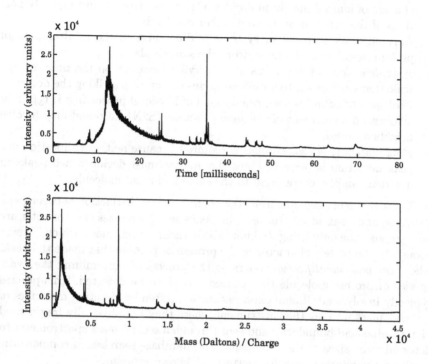

Fig. 4.2. A sample spectrum displayed on two scales. **(Top)** Intensity data as a function of the actual time-of-flight. **(Bottom)** Intensity as a function of the calibrated mass-to-charge ratio. Mass is measured in Daltons; charge is measured in multiples of the charge of one electron.

A typical data set arising in a clinical application of mass spectrometry contains tens or hundreds of spectra; each spectrum contains many thousands of intensity measurements representing an unknown number of protein peaks. Any attempt to make sense of this volume of data requires extensive low-level processing in order to identify the locations of peaks and to quantify their sizes accurately. Inadequate or incorrect pre-processing methods, however, can result in data sets that exhibit substantial biases and make it difficult to reach meaningful biological conclusions (Baggerly et al., 2003; Sorace and Zhan, 2003; Baggerly et al., 2004b,a). The low-level processing of mass spectra involves a number of complicated steps that interact in complex ways. Typical processing steps are as follows.

82   Kevin R. Coombes, Keith A. Baggerly, and Jeffrey S. Morris

- *Calibration* maps the observed time of flight to the inferred mass-to-charge ratio.
- *Filtering* or *denoising* removes random noise, typically electronic or chemical in origin.
- *Baseline subtraction* removes systematic artifacts, usually attributed to clusters of ionized matrix molecules hitting the detector during early portions of the experiment, or to detector overload.
- *Normalization* corrects for systematic differences in the total amount of protein desorbed and ionized from the sample plate.
- *Peak detection* is the process of identifying locations on the time or m/z scale that correspond to specific proteins or peptides striking the detector.
- *Peak quantification* is the primary goal of low-level processing; it typically involves an assessment of the signal-to-noise (S2N) ratio and may involve heights or areas.
- *Peak matching* across samples is required because neither calibration nor peak detection is perfect. Thus, the analyst must decide which peaks in different samples correspond to the same biological molecule.

In the realm of mass spectrometry, there is a clear distinction between peak *detection* and peak *identification*. The peaks seen by a mass spectrometer are anonymous. The only thing we know about them is their mass, which is never enough to completely characterize the protein or peptide that made the peak. The term *peak identification* refers to the process of determining the exact species of protein molecule that caused a peak to be detected. This process typically involves additional experimentation (often by shunting molecules of a target mass into another instrument where they are physically fragmented along amino acid boundaries and sent through a second mass spectrometer to determine the sizes of the fragments) and database searches to compare the results with the fragmentation patterns of known proteins.

The potential importance of the clinical applications of mass spectrometry has drawn the attention of increasing numbers of analysts. As a result, the development of better methods for processing and analyzing the data has become an active area of research (Rai et al., 2002; Baggerly et al., 2003; Coombes et al., 2003; Hawkins et al., 2003; Lee et al., 2003; Liggett et al., 2003; Wagner et al., 2003; Yasui et al., 2003a,b; Zhu et al., 2003; Coombes et al., 2005b; Morris et al., 2005). One should note that not all methods use all of the processing steps listed above, nor do they necessarily perform them in the same order.

4.2 Basic Concepts

Statistically, the low-level processing of mass spectra reduces to decomposing the observed signal into three components: True signal, baseline, and noise. One might try to decompose a spectrum using a model represented schematically by the equation

$$f(t) = B(t) + N \cdot S(t) + \epsilon(t) \tag{4.1}$$

where $f(t)$ is the observed signal, $B(t)$ is the baseline, $S(t)$ is the true signal, N is a normalization factor, and $\epsilon(t)$ is the noise. At present, this model is of limited utility, since we do not have an effective characterization of the individual components. The true signal can, in principle, be modeled as a sum of independent, possibly overlapping, peaks, each corresponding to a single protein. Approximate shapes of the peaks might be estimated empirically by simulating the physical process by which a time-of-flight (TOF) mass spectrometer collects data (Coombes et al., 2005a; Morris et al., 2005). White noise is a plausible model for the final term in the model, based on the notion that it arises primarily from electronic noise in the detector. One might also argue that at least some components of the noise have additional structure that is time dependent or even periodic (Baggerly et al., 2003). A fundamental limitation of the model in Equation 4.1, however, is that we do not have a good theoretical model for the baseline, aside from the vague intuition that it consists of a very low frequency component of the observed signal. This intuition is difficult to use without making it more precise, because the shape of the true peaks changes within a spectrum, becoming significantly lower and broader at later times and higher masses.

Our current procedure for processing sets of mass spectra is founded on two principles. First, the raw data is the ultimate arbiter; processing should be kept to a minimum in order to avoid introducing additional variance or additional bias into the measurements that will be used in later statistical analyses. Second, we should borrow strength across samples whenever possible.

1. Align the spectra on the time scale by choosing a linear change of variables for each spectrum in order to maximize the correlation between spectra.
2. Compute the mean of the aligned raw spectra.
3. Denoise the mean spectrum using the *undecimated discrete wavelet transform* (UDWT).
4. Locate intervals containing peaks by finding local maxima and minima in the denoised mean spectrum.
5. Quantify peaks in individual raw spectra by recording the difference between the maximum height and minimum height in each interval that should contain a peak.
6. Calibrate all spectra using the mean of the full set of available calibration experiments.

4.3 Advantages and Disadvantages

The chief advantage of performing peak finding by locating intervals in the mean spectrum that contain peaks is that it avoids the extremely messy and error-prone problem of matching peaks across spectra. The corresponding disadvantage is that this will only work if the spectra have been aligned properly before computing the mean (Figure 4.3). A small amount of misalignment is safe; it merely broadens the peaks in the mean spectrum. Severe misalignment, however, can make the data unusable.

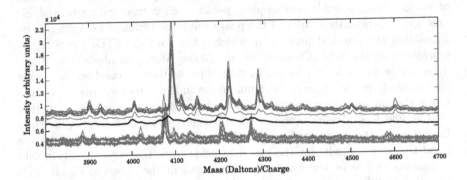

Fig. 4.3. Mean spectrum on improperly aligned data. The same sample was processed in multiple laboratories for several weeks. The two sets of gray curves are spectra from different laboratory-weeks. The heavy black curve is the mean spectrum over all laboratories and weeks. The sharp peaks that are present in the individual spectra have been diluted in the mean spectrum by a failure to align the spectra properly.

There are two advantages that follow from performing alignment on the time scale rather than first calibrating and then aligning on the mass scale. First, it is simpler, since it only requires a linear change of variables instead of a quadratic. Second, it is more reproducible, since it does not incorporate any additional errors that might be introduced in the calibration step. This factor is particularly important in many of the applications of mass spectrometry to protein profiling of complex mixtures. In many studies, the instrument is only calibrated in a fairly narrow range, but data is collected over a much wider range. For example, Ciphergen has a low mass standard mixture that contains five proteins with masses between 1084.2 and 7033.6 Daltons; their high mass standard mixture contains proteins with masses between 12.2 and 116.4 kiloDaltons. Both calibrant mixtures have been used while acquiring spectra from 1 000 to 50 000 Daltons or higher. When the calibration is extrapolated in this way, the errors can be substantial. Our final calibration step, which averages the results of multiple calibration experiments, should perform more accurately, even when extrapolated, than using a single calibration experiment.

The peak quantification step in our procedure implicitly performs local baseline correction without fitting an explicit curve. The local minimum in the interval containing the peak is taken to be the local definition of baseline. Without a coherent model that explicitly describes the shape baseline takes, preferably one motivated by the physical processes that affect the detector in a mass spectrometer (Malyarenko et al., 2005), fitting baseline can be problematic. Using the local minimum as an estimate of baseline has several advantages. First, it is simple to compute. Second, it does not require fitting either a parametric or nonparametric model that may simply not be appropriate in some circumstances. For example, the spectrum in Figure 4.2 has a baseline that might be modeled by an exponential decay starting at a high point near 12 ms. The baseline before 12 ms, however, clearly has a different shape. We have also seen spectra with two large bumps instead of one, which makes it difficult to specify a model that will work in full generality.

Another advantage of quantifying the peak height as the difference between local maximum and local minimum on a nonempty interval is that it avoids assigning a quantification of zero. Nonexistent peaks in a sample will be assigned a value that is proportional to the noise in the spectrum. By biasing the estimates slightly high in this manner, it is easier to work with transformations of the peak height in later statistical analyses of the data. When using alternative methods that assign a value of 0, analysts who want to use a log-transformation typically make an arbitrary choice to truncate the data before transformation. In essence, our method accepts additional bias in order to reduce some of the variance and avoid depending on arbitrary thresholds.

A critical disadvantage, however, is that the height of overlapping peaks can be biased significantly low (Figure 4.4). If a peak overlaps with other peaks on both sides, then the local minima will not come all the way down to the true baseline. In many cases, such overlapping peaks often represent related molecules that will be highly correlated in expression. There are a number of phenomena that give rise to such related molecules. For example, some proteins can carry along one or more matrix molecules (or *adducts*). The acids used in the matrix typically have a mass between 100 and 200 Daltons. A collection of regularly spaced peaks with mass difference in this range often represents the same protein or peptide carrying different numbers of matrix adducts. Proteins can also pick up sodium ions (changing mass by 22 Daltons) or lose a water molecule (with a mass of 18 Daltons). So, peaks whose mass difference is 18 or 22 Daltons also often represent the same protein or peptide. At a finer scale, isotopes of carbon (^{12}C vs. ^{13}C), nitrogen (^{14}N vs. ^{15}N), oxygen (^{16}O vs. ^{18}O) or other common elements can be incorporated into proteins in different numbers, leading to chemically identical proteins that differ in mass by 1 or 2 Daltons. Most mass spectrometers can be focused, at least at low mass levels, to be able to resolve differences smaller than a single Dalton, which occur when ionized proteins acquire multiple charges.

The mass spectrometry community appears to be converging on the use of wavelets for denoising. Because the intrinsic shape of a peak changes with the mass (becoming broader and lower at higher mass), the adaptive, multiscale nature of wavelets makes them a natural choice for denoising mass spectra, since these properties allow them to efficiently capture peaks of different

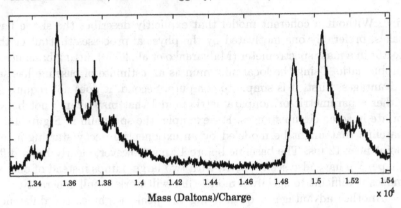

Fig. 4.4. Closeup of a raw spectrum. The two peaks indicated by arrows overlap with the peaks on either side, so the local minima closest to these peaks do not go all the way down to baseline.

widths. The wavelet approach for denoising involves three steps. The first is to compute the wavelet coefficients from the data, which involves choosing a basic wavelet basis function, then applying a series of linear filters derived from this function in a pyramid-based algorithm, called the *discrete wavelet transform* (Mallat, 1989). Applying this transform to a set of spectra results in a vector of wavelet coefficients summarizing signals at different frequencies and locations within the spectra. Second, set small wavelet coefficients to zero (*thresholding*), and third, compute the inverse wavelet transform to recover the denoised spectrum. The larger coefficients not set to zero can either be shrunken towards zero (*soft thresholding*) or left as they are (*hard thresholding*). In our experience, hard thresholding seems to perform better in denoising applications, since it results in less bias in the reconstructed denoised signal. Researchers still have a number of choices to make when using wavelets, however. They must select a basic wavelet basis function on which to base the transform (we usually use a Daubechies wavelet of degree 8, (Daubechies, 1992), the kind of transform (we use the UDWT (Lang et al., 1995, 1996; Gyaourova et al., 2002)), and the thresholding procedure (we use hard thresholding, with the threshold determined manually). The UDWT is superior to the more common decimated discrete wavelet transform (DDWT) when it comes to denoising. Its primary advantage is that, by construction, the UDWT is shift-invariant. The DDWT, by contrast, can produce different results if the start of the signal is shifted by a few time points. As a consequence, denoising with the DDWT can introduce significant artifacts into the signal near either end of the spectrum.

4.4 Caveats and Pitfalls

We have already mentioned some of the major difficulties that can arise using this procedure. First, the spectra must be properly aligned on the time scale. If this step is not performed correctly, then the peaks can be completely "out of phase" in some regions of the spectra, causing them to disappear from the mean spectrum. One also has a choice of trying to compute all pairwise alignments or just selecting a "standard" spectrum and aligning all other spectra with the standard. Using all pairwise alignments can lead to computationally challenging optimization problems. By contrast, the alignments can potentially vary if one standard spectrum is replaced with another. Our own practice is to use the "most typical" spectrum as a standard to which all others are aligned. In order to select the most typical spectrum, we first compute the mean spectrum without any alignment, and compute the Pearson correlation between this unaligned mean and each spectrum. The most typical spectrum is defined to be the one that maximizes the correlation with the mean.

One concern is that protein peaks that are present in only a few spectra will not be detectable in the mean. In an extensive simulation study, we compared peak finding using the mean spectrum to peak finding in individual spectra followed by matching peaks across spectra (Morris et al., 2005). Large peaks, even if rare, can still be found in the mean. Peaks that are small and rare are harder to find, but our simulations indicate, as a reasonable rule of thumb, that any peak that is present in at least $\sqrt{N}$ spectra, where N is the number of spectra in the study, is as likely to be detected in the mean as it is in individual spectra. If you believe that it is important to find small peaks that are present in fewer than $\sqrt{N}$ spectra, than you will have to supplement the mean spectrum approach with the study of individual spectra. In the situation where there are natural biological groups of spectra (for example, cancer patients vs. healthy controls), one may be able to restrict peak finding to the group mean spectra and the overall mean. In this approach, the peaks in the overall mean would be used to match most of the peaks found in the group means, and rare peaks that are present in only one group could still be located.

Our preliminary studies using the UDWT suggest that the degree of the Daubechies wavelet does not affect the results very much, so it is probably safe to use the one of degree 8 (Coombes et al., 2005b). Using hard thresholding also appears to do a better job than soft thresholding of preserving the actual shape of peaks. The only problematic part of wavelet denoising is selecting the threshold at which to truncate the wavelet coefficients. We use a variant of a SiZer plot (Chaudhuri and Marron, 1999) to select a threshold interactively. Our SiZer routine computes the denoised spectra over a user-specified range of thresholds, including one extreme value that provides a "super-smooth" curve. The differences between the super-smooth curve and the various denoised spectra are displayed in a heatmap, with time along the horizontal axis and

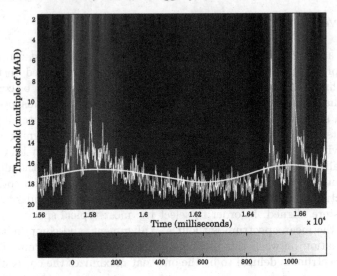

Fig. 4.5. SiZer plot of the effect of different wavelet thresholds (vertical axis) on the deviations of denoised spectra from a highly smoothed version (white curve).

thresholds along the vertical axis. The raw spectrum and the super-smooth curve are overlaid on top of the heatmap. In the example in Figure 4.5, most of the noise has been removed by the time the threshold reaches 4 or 5. The rightmost of the set of three peaks centered around 15 800 clock ticks appears to fade by the time the threshold reaches about 10 or 12. For this spectrum, a threshold between 6 and 10 looks appropriate. By focusing the SiZer plot on different regions of the spectrum, the analyst can refine this estimate and select a threshold that retains most of the visible peaks without following all the zigs and zags in the noise. It would, of course, be extremely useful if the selection of the threshold could be automated, preferably by defining a reasonable objective function of the threshold that could be optimized.

We have also described the biases that can occur in the heights of peaks that overlap their neighbors. One can, of course, insert any preferred baseline correction method between Steps 4 and 5 of the procedure described above. One would then have a choice of quantification methods available, including the maximum peak height or the area under the curve. Regardless of which method is used, however, a critical issue affecting downstream analysis of the resulting peak quantification matrix is the high level of correlation between peaks. Many successful analyses of mRNA expression microarray data have been conducted that either explicitly or implicitly assume that genes are independent. We suspect that the success of these methods has depended, at least in part, on the fact that the correlation matrix for gene expression is relatively sparse. The correlation matrix for protein peaks, by contrast, appears to be much denser. In addition to matrix adducts, sodium adducts, and isotope distributions that give rise locally to correlated peaks, there can also be distant correlation arising from the same protein present in the mixture in different charge states. (Keep in mind that we can only infer the mass-to-charge ratio from the time-of-flight, and cannot isolate the mass.) In some cases, there can be significant negative correlation between peaks that is both biologically and statistically significant. For example, phosphorylating a protein adds an 80-Dalton phosphate group to the unmodified protein, producing two peaks separated by 80 Daltons. Biologically, phosphorylation typically activates a protein, changing its behavior within the cell. It is certainly conceivable that one important difference between cancer cells and their healthy counterparts may lie not in the amount of a particular protein that is present but on the extent to which that protein is activated. If this is the case, then it could give rise to a pair of negatively correlated peaks separated in mass by 80 Daltons. In general, analysts dealing with peak quantification data from mass spectrometry experiments should be prepared to incorporate the correlation structure into their models.

The method described here does not perform normalization as a routine part of pre-processing. Analysts can still perform normalization later using the quantified peak heights. Such normalization can borrow techniques from the world of mRNA microarrays. For example, global normalization by dividing by the median peak height is likely to be robust and reasonably effective. One can also use linear mixed models in the spirit of Kerr et al. (2000) or Wolfinger et al. (2001) to incorporate peak-based normalization into the analysis of differential expression. Other alternatives for normalization are described in the next section.

4.5 Alternatives

Most alternative methods normalize by dividing by the *total ion current* (TIC), which is just the sum of the intensities under all or a substantial portion of the curve. Methods for computing TIC vary widely; it can be computed on raw data, baseline corrected data, or smoothed data. It can also be computed on the time scale or on the m/z scale. One must be careful on the m/z scale because some computations fail to account for the fact that the observations are no longer equally spaced. The total area under a curve estimated at a few thousand time points can be quite large; consequently, the normalized values are often multiplied by a large (arbitrary) constant to put the intensity units on a scale that doesn't require quite so many decimal points to display.

A basic suite of methods for processing SELDI data is implemented in the ProteinChip software from Ciphergen (Fung and Enderwick, 2002); these methods are comparable to those that have traditionally been used in the mass spectrometry community. Their default analysis is close to the order in our initial description of processing steps. They process one spectrum at a time, beginning with calibration to map the time-of-flight data to m/z values. They then perform baseline correction by fitting a varying-width segmented convex hull to the spectrum. Optionally, one can first smooth the spectrum by computing a moving average in a fixed width window before fitting the convex hull. Our own experience with Ciphergen's baseline correction suggests that it has a tendency to slice through the bottoms of peaks in areas of rapidly changing baseline (such as the region from 10 to 20 ms in Figure 4.2). They next denoise the spectrum either using a moving average or a Savitzky-Golay filter. The window size for the moving average can be constant on either the time scale or the m/z scale, or can vary over segments of the spectrum to account for the differences in the expected width of peaks. Their peak detection algorithm attempts to identify regions that rise above local valleys by a user-specified multiple of the noise. Peaks can be filtered based on the signal-to-noise ratio (S2N), whether the width of the peak at half-height is a specified multiple of the expected peak width, or by requiring the peak to have some minimum area. Normalization is performed by dividing by TIC or by the height or area of a specified control peak. Because the Ciphergen algorithm finds peaks in individual spectra, they must make a second pass to decide which peaks "match", or represent the same protein, in different spectra. They typically match peaks if their relative mass differs by a fixed percentage; this algorithm is based on the idea that the instrument has a nominal mass accuracy typically on the order of $0.1\% - 0.3\%$ across the entire range. In practice, such accuracies are probably achievable in the calibrated region, but the errors can be much larger when the calibrations are extrapolated to a wider range.

Yasui and colleagues (2003b) have described a method that does not attempt to quantify peaks; instead, they compute a binary indicator for the presence or absence of a peak. They define a point on the graph of the spectrum to be a peak if it satisfies two properties. First, it must be a local maximum in a fixed width window. (They use a window that extends 20 clock ticks on either side.) Second, it must have an intensity value higher than the average intensity in a broad neighborhood, where this average is computed using the super-smoother method in a window containing 5% of the data points. Because their downstream analysis only depends on presence or absence of peaks, they do not need to concern themselves with baseline correction, and denoising is implicitly accounted for by the super-smoother. They must still find an appropriate way to match peaks across spectra.

Our own pre-processing methods have evolved over time. Initially, we used a series of steps closely related to the Ciphergen routines (Baggerly et al., 2003). This method worked on calibrated spectra one at a time. We started by performing baseline subtraction using a "semi-monotonic" local baseline. We began by computing the local minimum in a fixed sized window (200 time steps). We next imposed a monotonicity requirement. (Note that this method would only make sense for the spectrum in Figure 4.2 by discarding the portion to the left of about 12 ms.) Since the combination of monotonicity with local minima would tend to be biased low as we moved to the right (and thus had a greater opportunity to see extremely low values of the noise), we added a "fuzz" parameter and computed the baseline as the smaller of the "monotone minimum + fuzz" and the "local minimum". We then normalized to TIC. The spectrum was then divided into windows whose width increased smoothly (along a quartic polynomial) across the spectrum. We quantified peaks as the maximum value in the baseline corrected spectrum in each window.

Our second method also worked on calibrated spectra one at a time (Coombes et al., 2003). This method performed peak-finding on the raw spectra, without baseline correction or denoising. Using first differences, a large list of candidate peaks was generated from all local maxima in the raw spectrum. The median absolute value of the first differences was used as an estimate of noise, and any local maximum that did not rise above the nearest local minimum by more than the noise was eliminated. Next, local maxima that were separated by fewer than $T = 3$ time steps of $M = 0.05\%$ relative mass units were combined into a single maximum. Then any peak where the slope from the maximum down to the nearby local minima was less than half the noise was eliminated. After this preliminary peak list was generated, the intervals containing the peaks were removed from the spectrum and replaced by linear interpolations. The baseline was estimated from the peak-free spectra by taking the local minimum in a fixed width window. The process of peak-finding and removal for baseline estimation was iterated to produce a stable baseline-corrected spectrum with an associated peak list. Peaks were matched across spectra if they differed in time by T time steps or in relative mass by M units.

Our third method initially worked one spectrum at a time on calibrated spectra, but introduced the UDWT for wavelet denoising (Coombes et al., 2005b). Denoising was performed as the first step of processing, using hard thresholding as described above. Baseline correction used a monotone local minimum; normalization was performed by dividing by TIC. Peak finding was performed on the denoised, baseline-corrected, normalized spectrum. After wavelet denoising, every local maximum is a candidate peak. Since the wavelet transform also gives local estimates of the noise, the only filtering performed on the peaks was to remove candidate peaks with S2N below a threshold. Peaks were quantified by the height of the local maximum in the processed spectrum. Peaks were matched across spectra if they differed in location by at most $T = 7$ time steps or in relative mass by at most $M = 0.3\%$.

The next step in the evolution of our pre-processing routines was to introduce the idea of using the mean spectrum for preprocessing (Morris et al., 2005). In this approach, we first aligned the spectra and computed the mean. We then denoised the spectrum using the UDWT, baseline corrected with a monotone minimum, and found peaks in the mean spectrum by keeping all local maxima with S2N > 5. In order to quantify these peaks in the individual spectra, the spectra were also wavelet denoised, baseline-corrected using the monotone minimum, and normalized to TIC. The size of a peak in an individual spectrum was taken to be the maximum value of the processed spectrum in the interval defining the peak.

All of these methods experience some difficulty with overlapping peaks, since the quantification for one peak will also contain possibly contaminating information from overlapping peaks. One approach for dealing with this problem is to model the spectra as a sum of peaks, with the peaks represented by some parametric form, and perform *deconvolution*. Ideally, this modeling and deconvolution should appropriately partition each intensity among all overlapping peaks. One example of this approach is given by (Clyde et al., 2006), in which the authors represent the peaks using a sum of Lévy processes. While potentially improving the quantifications, deconvolution also has the potential to introduce errors and extra variability to the process. There is a need for careful studies comparing methods involving deconvolution with those that do not.

Almost all methods in existing literature for analyzing mass spectrometry data involve first performing peak detection and quantification, then analyzing the peaks. An alternative approach is to model the mass spectra as functions, for example using functional mixed models (Morris et al., 2006). This approach has the potential to identify differentially expressed regions of the spectra that might be missed by peak detection algorithms, and also can automatically adjust for systematic effects due to nuisance factors, e.g., block effects, affecting both the intensities (y-axis) and locations (x-axis) of the peaks. Further study is necessary to compare the functional and peak-based approaches to determine the advantages and disadvantages of each.

4.6 Case Study: Experimental and Simulated Data Sets for Comparing Pre-Processing Methods

As you can tell from the previous section, a wide variety of methods have been proposed for pre-processing mass spectra. Not surprisingly, it can be difficult to determine which methods are better than others. The evolution of our own thought on the matter (described in painful detail above) has been guided by two kinds of data sets: Actual experimental data consisting of replicate spectra from the same sample, and a large set of simulated data.

Our collaborators have been willing to produce data sets containing numerous replicate spectra, obtained by processing aliquots of the same sample on different days and different chips. Specifically, samples of nipple aspirate fluid (NAF) were collected from women with unilateral breast cancer and from healthy women using methods that we have described elsewhere (Kuerer et al., 2004; Pawlik et al., 2005). Small amounts of the samples from all women in the study were pooled to produce a single quality control (QC) sample. The QC sample was divided into aliquots and stored at $-80°$C. In an initial experiment, the QC sample was processed on two spots of each of three different eight-spot ProteinChip arrays (Ciphergen Biosystems, Inc., Fremont, CA).

This procedure was repeated for four successive days, producing a total of 24 spectra from the same sample. In all subsequent experiments with biological samples of interest, two spots of each eight-spot ProteinChip array were used for the QC sample. Since 36 additional arrays were used, this produced 72 more replicate spectra from the same QC sample, collected over several months. This data set allows us to compare pre-processing methods by examining the extent to which they produce reproducible results on replicate spectra (Coombes et al., 2005b). Details on how these samples were used for QC have been described elsewhere (Coombes et al., 2003).

We analyzed the initial set of 24 QC samples using several different algorithms (Coombes et al., 2005b). Because all the samples were the same, our main concern was whether the processing methods could reproducibly find the same peaks. First, we applied our wavelet-denoising algorithm with a threshold of 10 to individual spectra, using the "monotone minimum" to correct baseline. This method detected, on average, about 211 local maxima per spectrum in the region above 950 Daltons/charge. Of these local maxima, about 158 per spectrum had S2N > 2 and about 96 had S2N > 10.

Next, we analyzed the same spectra using the algorithm in the Ciphergen ProteinChip software. With the default parameter settings, the Ciphergen algorithm found only 9 peaks per spectrum. When we increased the "peak sensitivity" setting to maximum, making no other changes, then the Ciphergen algorithm found only 41 peaks per spectrum. Thus, the wavelet denoising method consistently found more peaks than the Ciphergen algorithm.

One possible explanation of the difference between the algorithms is that the Ciphergen algorithm is more conservative than the wavelet-based algorithm, and thus only finds the tallest, most reliable peaks. If this were the case, then we would expect the Ciphergen algorithm to be more reproducible across spectra. In order to test this possibility, we matched peaks across spectra if they differed in time by fewer than 7 time steps or in relative mass by less than 0.3%. With these matching criteria, the wavelet-based method found a total of 174 distinct peaks and the Ciphergen algorithm (at maximum sensitivity) found a total of 149 distinct peaks. We plotted a histogram counting the number of times, in 24 samples, that the same peak was identified as present (Figure 4.6). We found that with the wavelet-based algorithm, 47 peaks were present in all 24 spectra, 83 peaks were found in at least 20 spectra, and 130 peaks were found in at least 10 spectra. With the Ciphergen algorithm, by contrast, only 6 peaks were present in all 24 spectra, and 47 of the 149 distinct peaks were present in only 1 spectrum. On this data set, the wavelet-based methods not only identified more total peaks, but it identified them more reproducibly.

We also analyzed the same spectra using the method described by Yasui and colleagues (2003b). We applied their method with a grid of parameter values, letting the window parameter range take on the value 10, 20, ..., 100 and the smoothing parameter take on the values 0.01, 0.02, 0.05, 0.07, 0.10, 0.15, and 0.20. For each combination of parameters, we computed the mean and

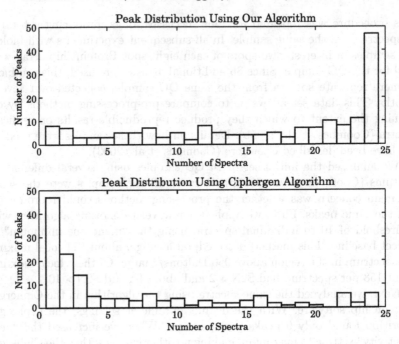

Fig. 4.6. Histograms showing the number of peaks found in replicate spectra. **(Top)** Our wavelet-based algorithm found 174 distinct peaks, and 47 of those peaks were found in all 24 spectra. **(Bottom)** The Ciphergen algorithm found 149 distinct peaks, but 47 of the peaks were identified in only one spectrum and only 6 peaks were identified in all 24 replicate spectra.

standard deviation of the number of peaks found in the 24 replicate spectra. The standard deviation was about the same (mean 64.26, range $60.36 - 70.43$) for all choices of the parameters. The mean number of peaks appeared relatively insensitive to the smoothing parameter, but decreased significantly as a function of the width parameter. Figure 4.7 shows a single spectrum in three different mass ranges. The overlaid curve is a super-smooth using 5% of the data points; circles indicate peaks found by Yasui's method using a window width of 80. With these parameters, their method detected an average of 267 "peaks" per spectrum. In the higher mass range (above 20 000 Da), these peaks do not appear to differ significantly from the surrounding noise. At lower mass ranges (between 2 000 and 3 000 Da), however, the window width prevented several clearly visible peaks from being detected. In the middle mass range, we also saw clear peaks (e.g, around 14 500 and 14 800 Daltons) that went undetected because they fell below the level of the super-smooth curve. If we decreased the window width or the super-smooth parameter in order to detect the obvious peaks in the low and middle mass ranges, we obtained vastly larger numbers of spurious peaks in the high mass region. The reproducibility across spectra of the peaks found by Yasui's method was comparable to those found by the Ciphergen algorithm (data not shown).

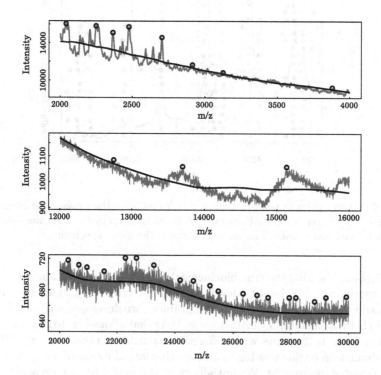

Fig. 4.7. Results of the peak-finding method proposed by Yasui and colleagues. The gray curve is the raw spectrum; the black curve is a super-smooth using 5% of the data. Circles mark local maxima that exceed the super-smooth level, which should correspond to peaks.

Reproducibility, by itself, is not enough to determine which method works better. One can potentially get more reproducible results by being very conservative about which features in a spectrum are called peaks. The largest peaks may be found very reproducibly, but the cost of a highly conservative approach is that a large number of smaller peaks may become "false negatives" — true peaks that cannot be used in later analyses because they were never found to begin with. Another potential problem is that the measure of reproducibility depends on matching peaks across spectra, using an algorithm that itself is not error-free. The matching step is required because even after calibration and alignment, peaks will not be perfectly aligned across replicates. Our matching algorithm joins peaks into "bins" if the difference in mass is less than 0.3%. Slight errors in alignment can combine with an occasional spurious peak to lump distinct peaks into a common bin (Figure 4.8).

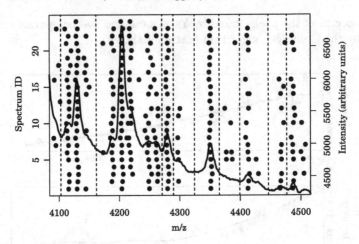

Fig. 4.8. Difficulties in peak matching. Circles indicate the presence or absence of peaks in the 24 replicate NAF spectra. Vertical lines mark the bins that separate distinct "matched" peaks. The overlaid curve is the mean spectrum.

Without knowing the true biochemical composition of the samples used in the experiments, it is hard to develop additional criteria by which to evaluate processing methods. To deal with this problem, we developed a simulation engine in S-Plus (Insightful Corp., Seattle, WA) that allowed us to simulate mass spectra from instruments with different properties (Coombes et al., 2005a). The simulation engine was based on a mathematical model of a physical mass spectrometry instrument. We initially used the model to explore some of the low-level characteristics of mass spectrometry data, including the limits on mass resolution and mass calibration, the role of isotope distributions, and the implications for methods of normalization and quantification. We then used the simulation engine to compare peak finding based on individual spectra to peak finding using the mean spectrum (Morris et al., 2005). We referred to the algorithm that matched peaks that were found by the wavelet-based algorithm on separate or single spectra as SUDWT. The algorithm that used the same denoising and baseline correction procedures but found peaks in the mean spectrum was called MUDWT.

For the simulation, we began with a *virtual population*, which is a distribution that describes the peaks that might be found in a *virtual sample* drawn from this population. An individual peak was characterized by four parameters: Its mass X, its mean M intensity on the log scale, its standard deviation S on the log scale, and its prevalence P, which is the probability that it is present in any given sample. We modeled the prevalence with a beta distribution and modeled the triple $(\log(X), M, S)$ with a multivariate normal distribution; the hyperparameters describing these distributions were estimated from real data. We simulated virtual populations containing 150

Table 4.1. Overall results from the simulation study. The top element in each box is the mean quantity over the 100 virtual experiments, and the bottom interval is the range. The comparison proportion p measures the proportion of the virtual experiments for which the MUDWT had higher sensitivity than the SUDWT plus one-half the proportion for which the methods tied.

Settings	Method	Sensitivity	FDR
$n=100$ $\sigma=66$	SUDWT	0.75 (0.60, 0.85)	0.09 (0.02, 0.26)
	MUDWT	0.83 (0.75, 0.92)	0.06 (0.00, 0.41)
	Comparison	0.97	0.80
$n=100$ $\sigma=22$	SUDWT	0.58 (0.43, 0.69)	0.25 (0.11, 0.41)
	MUDWT	0.74 (0.61, 0.84)	0.23 (0.10, 0.52)
	Comparison	1.00	0.63
$n=100$ $\sigma=200$	SUDWT	0.70 (0.61, 0.80)	0.08 (0.00, 0.17)
	MUDWT	0.78 (0.69, 0.87)	0.05 (0.00, 0.45)
	Comparison	0.97	0.86
$n=33$ $\sigma=66$	SUDWT	0.73 (0.63, 0.84)	0.09 (0.01, 0.20)
	MUDWT	0.80 (0.74, 0.86)	0.06 (0.00, 0.36)
	Comparison	0.99	0.85
$n=200$ $\sigma=66$	SUDWT	0.75 (0.58, 0.87)	0.12 (0.02, 0.46)
	MUDWT	0.85 (0.75, 0.91)	0.11 (0.00, 0.31)
	Comparison	1.00	0.69

peaks. In order to simulate a *virtual experiment*, we drew N samples from the population, processed them through our virtual mass spectrometer, and added Gaussian white noise with mean zero and standard deviation σ. For each combination of N and σ, we stimulated 100 different experiments. In each experiment, we applied both SUDWT and MUDWT to detect peaks. Performance of the algorithms was measured by the sensitivity (the proportion of true peaks matching at least one found peak) and the false discovery rate (FDR; the proportion of found peaks that matched no true peak). We found that, at comparable FDR levels, MUDWT had higher sensitivity overall than SUDWT (Table 4.1). SUDWT did have a slight advantage when detecting peaks at low abundance and low prevalence; see Morris et al. (2005) for details.

4.7 Lessons Learned

From our case study, we see that different pre-processing methods can lead to very different numbers of detected peaks. Thus, it is of crucial importance to identify approaches for comparing different methods and identifying which are most effective. We discussed two here. First, an experimental data set containing many replicate spectra from the same sample allows us to compare methods based on how reproducibly they detect peaks. Second, simulated spectra are useful for determining conditions under which different methods more accurately find and quantify peaks. We discussed a MALDI-TOF simulation engine that can be used to generate virtual spectra for which the true proteins and quantifications are known, and thus can be used to validate different methods. We focused on validating the peak detection step here, but it could be used equally well for comparing different denoising, baseline correction, and quantification methods, and could also be used to evaluate methods for identifying differentially expressed peaks and/or building classification models based on subsets of peaks.

4.8 List of Tools and Resources

Increased activity in the development of analytical tools to process mass spectra have produced a number of software packages.

1. A software package (Cromwell) implementing our methods in MATLAB (The MathWorks, Natick, MA) is available on our Web site at http://bioinformatics.mdanderson.org/software.html. The replicates in the NAF data set and the simulated data sets are also available by following the link to "Public Data Sets".
2. *Bioconductor* (http://www.bioconductor.org/), which began as a project to develop analysis tools in the statistical programming language R, has recently added a package called PROcess for the low-level processing of mass spectra.
3. The *Cancer Bioinformatics Grid* (caBig) is an effort by the United States National Cancer Institute to develop reusable software tools, standards, ontologies, and shared data. Progress of the caBig proteomics working group can be followed at the Web site https://cabig.nci.nih.gov/workspaces/ICR/Meetings/SIGs/Proteomics
4. Under the auspices of caBig, Duke University has been developing a suite of R programs to process mass spectra, called RProteomics (http://gforge.nci.nih.gov/projects/rproteomics).
5. The wavelet-based methods described in Coombes et al. (2005b); Morris et al. (2005) and the methods described in Yasui et al. (2003a,b) have been implemented as a commercial add-on, Proteome 1.0, to S-PLUS (Insightful, Seattle, WA).

6. Incogen (Williamsburg, VA), in cooperation with proteomics researchers at William and Mary College and the Eastern Virginia Medical School, has included support for the processing and analysis of mass spectra in its Visual Integrated Bioinformatics Environment (VIBE) software.

Naturally, manufacturers of mass spectrometers supply software with their instruments that does some form of basic pre-processing. When shifting away from the manufacturer's software to an alternative package, one has to worry about file formats. Ciphergen, for example, saves spectra in a proprietary binary format but also allows you to export them as comma-separated-values with two columns (m/z and intensity) or in a simple XML format. The XML file format is usually preferable, since it retains information about the protocol and the condition of the instrument when the spectrum was acquired. Two different efforts are underway to develop standard XML formats for mass spectrometry data. The de facto standard appears to be mzXML (described in detail at http://tools.proteomecenter.org/mzXMLschema.php), which is supported by conversion tools that accept the native format from several different MALDI-TOF instruments and was adopted by caBig. An alternative XML format, mzData (http://psidev.sourceforge.net/ms) is being developed by the Proteomics Standards Institute.

4.9 Conclusions

Numerous methods have now been suggested for pre-processing mass spectra, and both free and commercial software packages implementing these methods have become available. Because the methods can produce very different results, researchers interested in performing downstream analysis on the peak lists must make sure that the processing applied at the early stages is appropriate for their data. Ideas for quantifying which processing methods produce better results have started to be proposed, and data sets (both experimental and simulated) are available to start evaluating the performance of different methods. For most applications, it appears that peak detection using the mean spectrum is superior to methods that work with individual spectra and then match or bin peaks across spectra. Nevertheless, the development of better pre-processing methods remains an active area of research.

References

Adam, B.L., Qu, Y., Davis, J.W., Ward, M.D., Clements, M.A., Cazares, L.H., Semmes, O.J., Schellhammer, P.F., Yasui, Y., Feng, Z., and Wright, G.L., Jr. (2002). Serum protein fingerprinting coupled with a pattern-matching algorithm distinguishes prostate cancer from benign prostate hyperplasia and healthy men. *Cancer Res.*, 62(13):3609–3614.

Adam, P.J., Boyd, R., Tyson, K.L., Fletcher, G.C., Stamps, A., Hudson, L., Poyser, H.R., Redpath, N., Griffiths, M., Steers, G., Harris, A.L., Patel, S., Berry, J., Loader, J.A., Townsend, R.R., Daviet, L., Legrain, P., Parekh, R., and Terrett, J.A. (2003). Comprehensive proteomic analysis of breast cancer cell membranes reveals unique proteins with potential roles in clinical cancer. *J. Biol. Chem.*, 278(8):6482–6489.

Baggerly, K.A., Edmonson, S.R., Morris, J.S., and Coombes, K.R. (2004a). High-resolution serum proteomic patterns for ovarian cancer detection. *Endocr. Relat. Cancer*, 11(4):583–584; author reply 585–587.

Baggerly, K.A., Morris, J.S., and Coombes, K.R. (2004b). Reproducibility of SELDI-TOF protein patterns in serum: comparing datasets from different experiments. *Bioinformatics*, 20(5):777–785.

Baggerly, K.A., Morris, J.S., Wang, J., Gold, D., Xiao, L.C., and Coombes, K.R. (2003). A comprehensive approach to the analysis of matrix-assisted laser desorption/ionization-time of flight proteomics spectra from serum samples. *Proteomics*, 3(9):1667–1672.

Chaudhuri, P. and Marron, S. (1999). SiZer for exploration of structures of curves. *JASA*, 94:807–823.

Coombes, K.R., Fritsche, H.A., Jr., Clarke, C., Chen, J.N., Baggerly, K.A., Morris, J.S., Xiao, L.C., Hung, M.C., and Kuerer, H.M. (2003). Quality control and peak finding for proteomics data collected from nipple aspirate fluid by surface-enhanced laser desorption and ionization. *Clin. Chem.*, 49(10):1615–1623.

Coombes, K.R., Koomen, J.M., Baggerly, K.A., Morris, J.S., and Kobayashi, R. (2005a). Understanding the characteristics of mass spectrometry data through the use of simulation. *Cancer Informatics*, 1:41–52.

Coombes, K.R., Tsavachidis, S., Morris, J.S., Baggerly, K.A., Hung, M.C., and Kuerer, H.M. (2005b). Improved peak detection and quantification of mass spectrometry data acquired from surface-enhanced laser desorption and ionization by denoising spectra with the undecimated discrete wavelet transform. *Proteomics*, 5(16):4107–4117.

Clyde, M.A., House, L.L., and Wolpert, R.L. (2006). Nonparametric models for proteomic peak identification and quantification. *ISDS Discussion Paper*, 2006–2007.

Daubechies, I. (1992). Ten Lectures on Wavelets Philadelphia: Society for Industrial and Applied Mathematics.

Fung, E.T. and Enderwick, C. (2002). ProteinChip clinical proteomics: computational challenges and solutions. *Biotechniques*, Suppl.32, S34–S41.

Gyaourova, A., Kamath, C., and Fodor, I.K. (2002). Undecimated wavelet transforms for image de-noising. Technical Report UCRL-ID-150931, Lawrence Livermore National Laboratory, Livermore, CA.

Hawkins, D.M., Wolfinger, R.D., Liu, L., and Young, S.S. (2003). Exploring blood spectra for signs of ovarian cancer. *Chance*, 16:19–23.

Kerr, M.K., Martin, M., and Churchill, G.A. (2000). Analysis of variance for gene expression microarray data. *J. Comp. Biol.*, 7(6):819–837.

Kuerer, H.M., Coombes, K.R., Chen, J.N., Xiao, L., Clarke, C., Fritsche, H., Krishnamurthy, S., Marcy, S., Hung, M.C., and Hunt, K.K. (2004). Association between ductal fluid proteomic expression profiles and the presence of lymph node metastases in women with breast cancer. *Surgery*, 136(5):1061–1069.

Lang, M., Guo, H., Odegard, J. E., Burrus, C.S., and Well, R. O. Jr. (1995). Nonlinear processing of a shift invariant DWT for noise reduction. In Szu, H.H., editor, *Proc. SPIE. Waveelet Applications II*, volume 2491, pages 640–651, Bellingham, WA. SPIE.

Lang, M., Guo, H., Odegard, J.E., Burrus, C.S., and Well, R.O. Jr. (1996). Noise reduction using an undecimated discrete wavelet transform. *IEEE Signal Processing Letters*, 3:10–12.

Lee, K.R., Lin, X., and Park, D.C., and Eslava, S. (2003). Megavariate data analysis of mass spectrometric proteomics data using latent variable projection method. *Proteomics*, 3:1680–1686.

Liggett, W., Cazares, L., and Semmes, O.J. (2003). A look at mass spectral measurement. *Chance*, 16:24–28.

Mallat, S.G.(1989). A Theory for Multiresolution Signal Decompsition: The Wavelet Representation. *IEEE Trans. Patter Analysis and Machine Intelligence*, 11:674–693.

Malyarenko, D.I., Cooke, W.E., Adam, B.L., Malik, G., Chen, H., Tracy, E. R., Trosset, M.W., Sasinowski, M., Semmes, O.J., and Manos, D.M. (2005). Enhancement of sensitivity and resolution of surface-enhanced laser desorption/ionization time-of-flight mass spectrometric records for serum peptides using time-series analysis techniques. *Clin. Chem.*, 51(1):65–74.

Merchant, M. and Weinberger, S.R. (2000). Recent advancements in surface-enhanced laser desorption/ionization-time of flight-mass spectrometry. *Electrophoresis*, 21(6):1164–1177.

Morris, J.S., Coombes, K.R., Koomen, J., Baggerly, K.A., and Kobayashi, R. (2005). Feature extraction and quantification for mass spectrometry in biomedical applications using the mean spectrum. *Bioinformatics*, 21(9):1764–1775.

Morris, J.S., Brown, P.J., Herrick, R.H., Baggerly, K.A., and Coombes, K.R. (2006). Bayesian analysis of mass spectrometry proteomics data using wavelet based functional mixed models. *UT MD Anderson Cancer Center Department of Biostatistics and Applied Mathematics Working Papers Series*, Working Paper 22:1–32.

Paweletz, C.P., Trock, B., Pennanen, M., Tsangaris, T., Magnant, C., Liotta, L.A., and Petricoin, E.F., 3rd (2001). Proteomic patterns of nipple aspirate fluids obtained by SELDI-TOF: Potential for new biomarkers to aid in the diagnosis of breast cancer. *Dis. Markers*, 17(4):301–307.

Pawlik, T.M., Fritsche, H., Coombes, K.R., Xiao, L., Krishnamurthy, S., Hunt, K.K., Pusztai, L., Chen, J.N., Clarke, C.H., Arun, B., Hung, M.C., and Kuerer, H.M. (2005). Significant differences in nipple aspirate fluid protein expression between healthy women and those with breast cancer demonstrated by time-of-flight mass spectrometry. *Breast Cancer Res. Treat.*, 89(2):149–157.

Petricoin, E.F., 3rd, Ornstein, D.K., Paweletz, C.P., Ardekani, A., Hackett, P.S., Hitt, B.A., Velassco, A., Trucco, C., Wiegand, L., Wood, K., Simone, C.B., Levine, P.J., Linehan, W.M., Emmert-Buck, M.R., Steinberg, S.M., Kohn, E.C., and Liotta, L.A. (2002). Serum proteomic patterns for detection of prostate cancer. *J. Natl. Cancer Inst.*, 94(20):1576–1578.

Rai, A.J., Zhang, Z., Rosenzweig, J., Shih, I.M., Pham, T., Fung, E.T., Sokoll, L.J., and Chan, D.W. (2002). Proteomic approaches to tumor marker discovery. *Arch. Pathol. Lab. Med.*, 126(12):1518–1526.

Schaub, S., Wilkins, J., Weiler, T., Sangster, K., Rush, D., and Nickerson, P. (2004). Urine protein profiling with surface-enhanced laser-desorption/ionization time-of-flight mass spectrometry. *Kidney Int.*, 65(1):323–332.

Sorace, J.M. and Zhan, M. (2003). A data review and re-assessment of ovarian cancer serum proteomic profiling. *BMC Bioinformatics*, 4:24.

Tang, N., Tornatore, P., and Weinberger, S.R. (2004). Current developments in SELDI affinity technology. *Mass Spectrom. Rev.*, 23(1):34–44.

Wagner, M., Naik, D., and Pothen, A. (2003). Protocols for disease classification from mass spectrometry data. *Proteomics*, 3(9):1692–1698.

Wellmann, A., Wollscheid, V., Lu, H., Ma, Z.L., Albers, P., Schutze, K., Rohde, V., Behrens, P., Dreschers, S., Ko, Y., and Wernert, N. (2002). Analysis of microdissected prostate tissue with proteinchip arrays–a way to new insights into carcinogenesis and to diagnostic tools. *Int. J. Mol. Med.*, 9(4):341–347.

Wolfinger, R.D., Gibson, G., Wolfinger, E.D., Bennett, L., Hamadeh, H., Bushel, P., Afshari, C., and Paules, R.S. (2001). Assessing gene significance from cDNA microarray expression data via mixed models. *J. Comp. Biol.*, 8(6):625–637.

Yasui, Y., McLerran, D., Adam, B. L., Winget, M., Thornquist, M., and Feng, Z. (2003a). An automated peak identification/calibration procedure for high-dimensional protein measures from mass spectrometers. *J. Biomed. Biotechnol.*, 2003(4):242–248.

Yasui, Y., Pepe, M., Thompson, M.L., Adam, B.L., Wright, G.L., Jr., Qu, Y., Potter, J.D., Winget, M., Thornquist, M., and Feng, Z. (2003b). A data-analytic strategy for protein biomarker discovery: profiling of high-dimensional proteomic data for cancer detection. *Biostatistics*, 4(3):449–463.

Zhu, W., Wang, X., Ma, Y., Rao, M., Glimm, J., and Kovach, J.S. (2003). Detection of cancer-specific markers amid massive mass spectral data. *Proc. Natl. Acad. Sci. USA*, 100(25):14666–14671.

Zhukov, T.A., Johanson, R.A., Cantor, A.B., Clark, R.A., and Tockman, M.S. (2003). Discovery of distinct protein profiles specific for lung tumors and pre-malignant lung lesions by SELDI mass spectrometry. *Lung Cancer*, 40(3):267–279.

5

Visualization in Genomics and Proteomics

Xiaochun Li[1,2] and Jaroslaw Harezlak[2]

[1] Dana Farber Cancer Institute, Boston, Massachusetts, USA.
   `xiaochun@jimmy.harvard.edu`
[2] Harvard School of Public Health, Boston, Massachusetts, USA.
   `jharezla@hsph.harvard.edu`

5.1 Introduction

In the age of high-throughput biological technology, experimental data have grown exponentially. Searching for data structures and succinctly presenting them is challenging but all the more essential. Genomic and proteomic data are by nature multi-dimensional. Different approaches and tools are needed for visualization to aid the exploration as well as quality assessment of the experimental data. The bon mot "A picture is worth a thousand words" expresses the importance of visualization in conveyance of information. Visualization tools not only help us to communicate but also help us to think, organize knowledge and discover new patterns. The indispensability of visualization is best attested by its extensive day-to-day use in presentations, papers and books. Recent discussion about ideas and tools pertaining to genomic and proteomic data can be found in (Gentleman et al., 2005; Hahne et al., 2006).

Data can be presented in various ways, as raw data, after some summarization or after analysis, each for different purposes. Snapshot-like images of an experimental plate, for example, false color image of 96 wells, or an Affymetrix chip, may be shown for a visual assessment of data quality. Wells are displayed in the actual physical locations and intensity values of wells may be represented either by consecutive gray levels or by certain color continuum. There may be patterns that are related to the positions of the wells rather than the underlying biological traits. One example is local image contamination, which human eyes are apt to detect. As another example, wells on the edges of chips or plates may have consistently higher or lower intensities than the wells in the middle because of different humidity, temperature or the sample amount. Inspection of array images can guide us in devising strategies for both analysis and experimental design. If edge effects are present, we may want to include array edges in a statistical model; or we may not want to use wells along array edges at all in future experiments. If a spatial pattern is present, we may want to use a Latin Square layout for placement of samples. A random or pseudo-random layout of samples is generally advised.

Ultimately, we are interested in relationships among variables, relationships among samples and relationships between variables and samples. Genomic or proteomic data, after necessary steps of pre-processing entailed by the specific technology used to generate the data, e.g., baseline subtraction and normalization for SELDI-TOF (surface-enhanced laser desorption/ionization time-of-flight) mass spectrometry data, are put in a $p \times n$ matrix where p is the number of genes or mass over charge ratios and n is the number of subjects or samples under various conditions. Note that the formulation of this matrix representation of objects is different from the conventional matrix representation of objects where rows are observations or records and columns are variables. In genomic or proteomic data the norm is $p \gg n$, the number of variables overwhelms the number of samples. This poses new methodological and computational challenges because classic statistical tools and theory deal with the opposite situation of $p < n$. The advancement of science and technology calls for new tools. An agglomeration of ideas and tools for visualization and analysis of genomic and proteomic data are presented in Gentleman et al. (2005). Various aspects of data can be graphically displayed for exploration. For example, if discovery and description of unknown subtypes of a disease is of interest, heatmaps (Eisen et al., 1998), now routinely used in bioinformatics, may help in visualizing patterns or structure in data from microarray or proteomic experiments. Typically, genes or proteins are arranged in rows and samples or patients are arranged in columns. Dendrograms of genes and samples may be added to the row or column margins to help organize the heatmap. Heatmaps, together with the dendrograms in the margins, are often useful as a data exploratory tool to see if results of clustering of samples coincide with clinical information, and if so what gene or protein/peptide patterns are associated with clusters of samples. Results of clustering of samples may also suggest new subtypes of disease and association between patterns of certain genes, proteins or pathways and potential new subtypes of disease.

More in-depth visualization can be done on genes in a particular chromosome for amplification or deletion. For example, the successive cumulative sums of gene expressions ordered by their physical locations on a chromosome can be displayed against the physical locations. By using such a visualizing technique it is hoped that if amplification in a single gene is too subtle, the successive aggregated signal of the whole region may be large enough to be visible. A nice example showing samples with trisomy 21 versus diploid samples can be found in chapter 10 of Gentleman et al. (2005).

Interested readers may want to peruse Gentleman et al. (2005) and Hahne et al. (2006) for general ideas about visualization and ideas pertaining to particular technologies. These ideas and approaches are either available directly from R, a language and environment for statistical computing and graphics (Ihaka and Gentleman, 1996), or from R-based software packages, which can be conveniently downloaded from the *Bioconductor* Web site, http://www.bioconductor.org/.

Other aspects of the experimental data, for example, pathways or networks can be represented as graphs (see Chapter 10). Models can be built and various statistics can be computed. We refer readers to Chapter 9, Gentleman et al. (2005) and Hahne et al. (2006) for detailed discussions.

In this chapter, we focus on one particular visualization technique, *multidimensional scaling* (MDS). Multidimensional scaling originated in psychometrics, and is also known as *perceptual mapping*. It is a method of visualizing the similarity of a set of objects, where each object is characterized by a collection of traits. MDS transforms similarities into distances in the familiar Euclidean space.

In the context of this chapter, a typical object is a numeric vector in a multidimensional space $\Re^p$. Often we want to explore the relationships among the objects. For example, a numeric vector can be the gene expression data of a patient sample using certain microarray technology, or a mass spectrum obtained from a serum sample of a patient using SELDI-TOF MS. The dimensionality of p in those circumstances is either the number of genes or the number of mass over charge ratios. Each vector of gene expressions from a microarray experiment, or relative abundance of peptides from a proteomic experiment can be represented by a point in $\Re^p$. Since p in general is very large and can be in the order of thousands, MDS may be used to achieve dimension reduction and to display the objects in a lower dimensional space $\Re^d$, $d \ll p$. Typically, $d = 2$ or 3 and object relationships can be easily visualized.

5.2 Basic Concepts

To explore the similarity among objects, we need firstly to define a measure of similarity, or alternatively, dissimilarity. Denote the dissimilarity between the ith and the jth objects as δ_{ij}. We will consider dissimilarity in the following sense,

1. The dissimilarity of an object with itself is zero, $\delta_{ii} = 0$;
2. The dissimilarity between two objects is non-negative, $\delta_{ij} \geq 0$;
3. The dissimilarity between two objects is symmetric, that is, $\delta_{ij} = \delta_{ji}$.

If dissimilarity δ_{ij} further satisfies the following properties:

4. $\delta_{ij} = 0$ if and only if $i = j$;
5. The triangle inequality holds, $\delta_{ij} \leq \delta_{ik} + \delta_{kj}$,

then the dissimilarity measure is called a *metric* (Mardia et al., 1979).

There are many dissimilarity measures to choose from to gauge the "unlikeness" between a pair of objects, or points. The most obvious measure of "unlikeness" in $\Re^p$ is the Euclidean distance (which is also a metric). Although some dissimilarity measures are metrics, within the context of this chapter, it is sufficient to consider dissimilarity measures that satisfy conditions 1-3.

Similarities between objects are given explicitly or are computed from a data matrix. For example, we may compute the Pearson correlation coefficients between pairs of patient samples to explore whether a gene or protein profile suggests a grouping of patients. The Pearson correlation coefficient is a *similarity* measure. We normally require that similarity s satisfies (a) symmetry, that is, $s_{ij} = s_{ji}$; and (b) $s_{ii} \geq s_{ji}$, an object is most similar to itself. Common transformations from a similarity measure to a dissimilarity measure are, $\delta_{ij} = c - s_{ij}$ for some constant $c \geq \max_{i,j} s_{ij}$, or $(s_{ii} - 2s_{ij} + s_{jj})^{1/2}$. The latter transforms similarity s_{ij} to the Euclidean distance (therefore, a metric) if the matrix formed by s_{ij} is positive semi-definite (Mardia et al., 1979). A popular choice of dissimilarity in genomic or proteomic studies is one minus the Pearson correlation coefficient (which is not a metric although its square root is).

In summary, pair-wise dissimilarities, if not directly given, are computed from a multivariate data matrix, resulting in a matrix of dissimilarities. Since a matrix of similarity is symmetric, there are $\frac{1}{2}n(n - 1)$ possibly distinct dissimilarities. The steps of an MDS analysis can be described in the diagram in Figure 5.1.

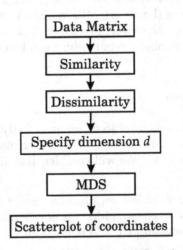

Fig. 5.1. Workflow of an MDS analysis.

Different forms of MDS exist that optimize different types of loss functions (or stress functions) and fall into categories of metric and nonmetric MDS. A loss function measures how close pairwise dissimilarities are preserved by a lower-dimensional approximation of the data. The distinction between metric and nonmetric MDS is whether the absolute values of dissimilarities or the *order* of dissimilarities are "matched" in some sense by distances (commonly

Euclidean) in a lower dimension $\Re^d$. Detailed discussions of loss functions and metric and nonmetric MDS can be found in Cox and Cox (2001).

Note that an MDS solution $\mathbf{Z}_{d\times n}$, is not unique in the sense that for any $d \times d$ orthogonal matrix $\mathbf{A}$ and an arbitrary vector $\mathbf{b}$ in $\Re^d$, $\mathbf{AZ} + \mathbf{b}$ is also a solution (Mardia et al., 1979). It can be easily verified that the Euclidean distances between the columns of $\mathbf{AZ}+\mathbf{b}$ are the same as the Euclidean distances between the columns of $\mathbf{Z}$. In other words, a low dimensional configuration produced by an MDS method is indeterminate with respect to translation, rotation, and reflection. This agrees with the intuition that the relationships between the data should be invariant to rigid motions of shifting, rotating and reflecting of the whole data cloud.

5.2.1 Metric Scaling

Metric MDS operates on a given matrix of dissimilarities $\mathbf{D} = \{\delta_{ij}\}$ or transformed dissimilarities, $f(\delta_{ij})$, where f is a *continuous* monotonic function, to find coordinates of a set of points in a lower dimensional space such that the distances are approximately preserved. In general, a perfect reproduction of dissimilarities, transformed or not, may not serve the purpose of dimension reduction and visualization. Even when distances can be reproduced, the number of dimensions may still be too large to be of any practical use. Hence loss functions that measure the "match" between original dissimilarities and distances in a lower dimensional $\Re^d$ are introduced and optimized.

Classical scaling is the original version of metric MDS, where the matrix of dissimilarities $\mathbf{D}$ is treated as Euclidean distances. Classical scaling minimizes the loss function of sum of differences of squared dissimilarities in the original space and squared distances in the reduced space $\Re^d$, that is, $\phi = \sum_{i=1}^{n}\sum_{j=1}^{n}(\delta_{ij}^2 - d_{ij}^2)$ (Mardia et al., 1979), where d_{ij} are Euclidean distances between objects in $\Re^d$.

Let $\mathbf{x}_i = (\mathbf{x}_{1i}, \ldots, \mathbf{x}_{pi})'$, $i = 1, \ldots, n$ be n points in $\Re^p$ and let the dissimilarity measure be the Euclidean distance. The *squared* Euclidean distance between the ith and j^{th} points is $\delta_{ij}^2 = (\mathbf{x}_i - \mathbf{x}_j)'(\mathbf{x}_i - \mathbf{x}_j)$. Starting from δ_{ij}^2, classical scaling finds the inner product matrix $\mathbf{B} = [b_{ij}]$ with $b_{ij} = \mathbf{x}_i'\mathbf{x}_j$ and from $\mathbf{B}$ the lower dimension coordinates.

Given a matrix of Euclidean distances $\mathbf{D}$, an algorithm of classical scaling is as follows:

1. Let $\mathbf{A} = [-\frac{1}{2}\delta_{ij}^2]$.
2. Find the inner product matrix $\mathbf{B}$ by doubly centering $\mathbf{A}$, $\mathbf{B} = \mathbf{HAH}$, where centering matrix $\mathbf{H} = \mathbf{I} - n^{-1}\mathbf{aa}'$, where $\mathbf{I}$ is the identity matrix and the n-dimensional vector $\mathbf{a} = (1, 1, \ldots 1)'$. $\mathbf{H}$ on the left centers the columns and $\mathbf{H}$ on the right centers the rows.
3. Find the eigenvalues of $\mathbf{B}$, $\lambda_1 \geq \lambda_2 \geq \ldots \geq \lambda_n$ and corresponding normalized eigenvectors $\mathbf{V} = (\mathbf{v}_1, \ldots, \mathbf{v}_n)$ with $\mathbf{v}_i'\mathbf{v}_i = 1$, $i = 1, \ldots, n$.

4. For a pre-specified d, the coordinates of the n points in $\Re^d$ are the rows of $(\sqrt{\lambda_1}\mathbf{v}_1, \ldots, \sqrt{\lambda_d}\mathbf{v}_d)$.

For detailed derivations, please see Cox and Cox (2001). Matrix $\mathbf{B}$ is not full rank due to centering and thus has at least one zero eigenvalue. Therefore in the above algorithm only $n - 1$ eigenvalues need to be sought. When $\mathbf{D}$ is Euclidean, or, equivalently, when $\mathbf{B}$ is positive semi-definite, all the eigenvalues are non-negative. In practice, we choose the configuration in $\Re^d$, often with $d \le 3$, whose coordinates correspond to the first d eigenvectors of $\mathbf{B}$.

It is well known that classical scaling, when the dissimilarities are Euclidean distances, is equivalent to principal component analysis using the sample covariance matrix (Cox and Cox, 2001). The connection can be sketched as follows. Without loss of generality, assume a data matrix $X_{p \times n}$ is row-centered since any matrix can be row-centered by multiplying the centering matrix $\mathbf{H}$ on its right. The inner product matrix can be written as $\mathbf{B} = \mathbf{X}'\mathbf{X} = \mathbf{V}\Lambda\mathbf{V}'$ (via spectral decomposition), where $\Lambda = diag(\lambda_1, \ldots, \lambda_n)$ and the columns of $\mathbf{V}$ are the corresponding normalized eigenvectors. The classical MDS solution in $\Re^d$ is $\mathbf{V}_1\Lambda_1^{1/2}$, where $\mathbf{V}_1$ is the first d columns of $\mathbf{V}$, and $\Lambda_1 = diag(\lambda_1, \ldots, \lambda_d)$. On the other hand, the covariance matrix from the same data matrix is $\mathbf{S} = (n-1)^{-1}\mathbf{X}\mathbf{X}'$. It can be shown that $\mathbf{X}'\mathbf{X}$ and $\mathbf{X}\mathbf{X}'$ have the same set of non-zero eigenvalues. Let ξ_i be the eigenvector corresponding to a non-zero λ_i, $\mathbf{X}\mathbf{X}'\xi_i = \lambda_i\xi_i$, with $\xi_i'\xi_i = 1$. Since $\mathbf{X}'\mathbf{X}\mathbf{v}_i = \lambda_i\mathbf{v}_i$, left-multiplying both sides by $\mathbf{X}$ gives $\mathbf{X}\mathbf{X}'\mathbf{X}\mathbf{v}_i = \lambda_i\mathbf{X}\mathbf{v}_i$ and thus $\xi_i = c_i\mathbf{X}\mathbf{v}_i$, where c_i is a constant. Since $1 = \xi_i'\xi_i = c_i^2\mathbf{v}_i'\mathbf{X}'\mathbf{X}\mathbf{v}_i = c_i^2\lambda_i\mathbf{v}_i'\mathbf{v}_i = c_i^2\lambda_i$, we have $c_i = \lambda_i^{-1/2}$. The first d principal component scores are

$$\mathbf{X}'[\xi_1, \ldots, \xi_d] = \mathbf{X}'[c_1\mathbf{X}\mathbf{v}_1, \ldots, c_d\mathbf{X}\mathbf{v}_d] \tag{5.1}$$

$$= [c_1\mathbf{X}'\mathbf{X}\mathbf{v}_1, \ldots, c_d\mathbf{X}'\mathbf{X}\mathbf{v}_d] \tag{5.2}$$

$$= [c_1\lambda_1\mathbf{v}_1, \ldots, c_d\lambda_d\mathbf{v}_d] \tag{5.3}$$

$$= [\sqrt{\lambda_1}\mathbf{v}_1, \ldots, \sqrt{\lambda_d}\mathbf{v}_d], \tag{5.4}$$

which is the classical MDS solution as stated in 4th step in the above algorithm.

The algorithm can be applied to a more general dissimilarity matrix $\mathbf{D}$ where $\mathbf{D}$ is not necessarily Euclidean. If some eigenvalues are negative, we ignore them and choose the first d positive eigenvalues and their corresponding eigenvectors. For details of optimality in this case, see Cox and Cox (2001).

It is generally accepted that large dissimilarities dictate the final MDS picture. Another algorithm developed by Sammon (Cox and Cox, 2001) minimizes a weighted loss function, known as *weighted stress*,

$$\sum_{i<j} \frac{(\delta_{ij} - d_{ij})^2}{\delta_{ij}} / \sum_{i<j} d_{ij}.$$

Weighted loss up-scales the small dissimilarities so that they have more influence on the final MDS configuration.

5.2.2 Nonmetric Scaling

Nonmetric MDS methods were developed to treat ordinal data, in which only the ranks of dissimilarities are meaningful, not the actual values. The transformation of dissimilarities f now only needs to satisfy the monotonicity constraint. Ranks are invariant under strict monotone increasing transformation. Nonmetric MDS methods attempt to preserve rank orders of dissimilarities by optimizing various forms of loss functions.

Although nonmetric MDS methods were developed for ordinal or qualitative data, they can be applied to quantitative data as well, for example, data vectors in $\Re^p$ of samples from microarray experiments. In the final pictorial representation, it is the order of the dissimilarities of data vectors that are approximated, not the actual values of dissimilarities.

One such algorithm was developed by Shepard and Kruskal in the 1960s (Shepard, 1962a,b; Kruskal, 1964a,b) to minimize a loss function, termed squared stress,

$$\sum_{i<j}(f(\delta_{ij}) - d_{ij})^2 / \sum_{i<j} d_{ij}^2,$$

which measures the disagreement of the resulting distances and the transformed dissimilarities.

There are many other algorithms of metric and nonmetric MDS, based on various transformations, loss functions and weighting. We refer readers to Cox and Cox (2001) for a thorough exposition. Although the global minimum is desired when the loss function is being minimized, in general most of the MDS algorithms suffer from the local minima problem. That is, an MDS algorithm stops at a local minimum (rather than the global minimum) of the loss function and thus results in a sub-optimal configuration. The only exception is classical scaling where a unique solution exists up to rotations, reflections and translations. For discussion on diagnosis of local minima, see Buja and Swayne (2002).

5.3 Advantages and Disadvantages

MDS is useful as an unsupervised machine learning tool. Given a dissimilarity matrix, MDS can produce a pictorial representation of objects under investigation. Often MDS results are displayed in $\Re^2$ or $\Re^3$. In $\Re^2$ more than two principal coordinates can be explored by simple pairwise plots of the first few principal coordinates. If shown in $\Re^3$, data representation can be manipulated through rotations, which allow a data analyst to examine object relationships from different directions to understand the entirety of the data. In the context of genomic and proteomic studies, MDS is used both in exploring sample relationships and gene or biomarker relationships. If latter, MDS is merely used as a graphical representation of biomarker relationships and is applied to a relatively homogenous group of samples. For example, Choe et al. (2003) used MDS plot to display gene correlations and conjectured how they may interact in terms of pathways.

Note that the input to an MDS analysis is a dissimilarity matrix, and the output are n points in a space of d dimensions. If starting with an $n \times p$ multivariate data matrix, one needs to decide on a dissimilarity measure and compute a matrix of dissimilarities. Although data reduction may be achieved, there are in general no meaningful interpretations of coordinates in the low dimensional representation of the objects. The lack of intuition of the low dimensional space where the objects are visualized is a drawback for this type of projection-based methods. Even in the case of classical scaling, it is the distances that are perfectly reproduced, not the original multivariate data matrix. The orientations of the axes from the MDS analysis are arbitrary and can be rotated in any direction. In fact, distances remain the same subject to translations, rotations and reflections. Quantitative comparisons of the spatial relationships across different scalings do not make sense because they are associated with different functions that transform dissimilarities. MDS is rarely used as a statistical tool that allows explicit modeling of the data for inference, but is used in practice as an exploratory data analysis tool. What we should look for are clusters of points or particular patterns in the resulting pictorial representation of the data.

In addition to the local minima problem mentioned earlier, MDS algorithms except the classical MDS are iterative, computationally intensive and may diverge. Starting from a classical solution may be a good strategy in some cases (Ripley, 1996).

Although principal component analysis and classical MDS yield the same results, the computation of the former is more intensive than the latter in the context of genomic and proteomic studies. When $p \gg n$, the computation of the principal component analysis is quadratic in p, while the computation of the classical MDS is linear in p. Some implementation of principal component analysis algorithm may require $n \geq p$ and therefore will not work for the case of $p \gg n$.

Lastly, both metric and nonmetric MDS require f, the transformation function of dissimilarities, to be monotonic (order preserving). However, if there exists some non-linear structure in the data, for example, data points exist in a low-dimensional manifold rather than a low-dimensional sub-space, MDS may not be able to retrieve such structures.

5.4 Caveats and Pitfalls

The pictorial representation of an MDS analysis in $\Re^2$ or $\Re^3$ may not reveal any structure in the data if a large proportion of variation in the data is not explained by the first two or three dimensions. Objects may appear to be close but can in fact be very far apart in the original space because of the remaining variation of the data in other dimensions. For example, if two points of $\Re^3$ have the same x and y but different z values, the projection of them on the x-y plane ($\Re^2$) results in the same point, indistinguishable because of the lack of the information about the third dimension. Data analysts should be cautious in the interpretation of MDS plots because of this. Viewing a solution from different angles helps in general. The simple graphics function **pairs** in R can be used to generate 2-D marginal views of MDS results. Its application will be illustrated in the case study in Section 5.6.

The decision about the number of dimensions to be used in visualization is arbitrary, however, a plot showing a goodness-of-fit measure against the number of dimensions may be helpful. In classical scaling, two possible goodness-of-fit measures are

$$\alpha_{1,d} = \sum_i^d \lambda_i / \sum_i^n |\lambda_i|, \tag{5.5}$$

$$\alpha_{2,d} = \sum_i^d \lambda_i^2 / \sum_i^n \lambda_i^2. \tag{5.6}$$

These two measures quantify the proportion of a dissimilarity matrix explained by the d-dimensional classical scaling solution (Mardia et al., 1979). Goodness-of-fit in general means how well a model fits a set of observations. Measures of goodness-of-fit typically summarize the discrepancy between observed values and the values expected under the model in question. In our context, the model is that data can be represented in a low-dimensional space, $\Re^d$. The dimension d should be chosen as small as possible as long as the goodness-of-it is not seriously impacted. For example, the point where the goodness-of-fit curve starts to flatten can be chosen as d, the dimension for the MDS solution.

As mentioned earlier, large dissimilarities have much more influence on an MDS result than small dissimilarities. Consequently, if objects appear close in an MDS picture, it does not necessarily imply that they are similar, but it rather means that there exists a large set of objects from which they are approximately equally dissimilar (Buja and Swayne, 2002). As discussed in Section 5.2, Sammon mapping is an attempt to explore structure contained in small dissimilarities. Another approach, "within-groups MDS", was proposed by Buja and Swayne (2002) to explore further structures after disjoint groups are discovered in the whole data set. They proposed to use MDS for the objects in the same group. This is equivalent to minimizing the stress function with only those dissimilarities of objects that belong to the same group.

The most frequently used dissimilarity measure is the Euclidean distance. However, Aggarwal et al. (2001) show both theoretically and empirically that the notion of proximity in high dimensions is better represented with the use of lower values of k in the L_k metrics. Thus, for instance the L_1-metric (Manhattan distance) is preferable to the L_2-metric (Euclidean distance). They also introduce distance measure L_k with $0 < k < 1$ and show empirically improvements in clustering algorithms and similarity searches.

Clustering of objects suggested by MDS might not be stable in the presence of noise and thus the robustness of clusters may be of interest. Krzanowski (2006) discusses the stability of the solutions obtained via MDS. He advocates the use of leave-one-out cross-validation in the sensitivity analysis of the classical scaling. The procedure involves performing $n + 1$ separate analyses, one for the full dissimilarity matrix and n for an omission of each combination of ith row and i^{th} column ($i \in \{1, \ldots, n\}$). The coordinates of the points for each analysis are obtained via a spectral decompositions of the $n + 1$ dissimilarity matrices. Assessment of stability is performed by plotting around each point the smallest hyper-sphere containing a percentage (e.g., 95%) of the $n + 1$ points obtained from the cross-validation procedure. Points with non-overlapping hyper-spheres indicate their membership in different clusters. If feature selection is carried out before MDS, it should be incorporated into the cross-validation procedure in the assessment of the robustness of the MDS result.

5.5 Alternatives

In genomic or proteomic studies, there are thousands of features (genes or proteins/peptides). Both biologically differentiating or non-differentiating features contribute to dissimilarity calculation. Non-differentiating features are just white noise, which may overwhelm the few differentiating features and cause the problem of indifferentiation (Buja and Swayne, 2002). Consequently we may not be able to see any clustering of objects in the low dimensional representation of the data. Filtering data before application of MDS may help enhance signal and reduce noise. A collection of filters can be found in the R package genefilter from the *Bioconductor* Web site.

Non-specific filters that do not use sample class labels can be used. For example, coefficients of variation (cv) of features can be computed; and only those features with cv above a certain threshold are included in an MDS analysis. Specific filters, such as t-test to select features that are different between biological groups, can also be used. However, the MDS results obtained by using only the subset of differentiating features should be interpreted with caution. It is expected that we see in the picture that there are distinct groups, since the features are chosen to be the most differentiating between groups. We present here a possible way of resolving this circular reasoning. A data set can be randomly split into training and test sets. Features are selected in the training set and then used as input variables to an MDS analysis in the test set. The MDS result of the test set is then compared to that of the training set to see if the same pattern is present in the test set. This process can be carried out multiple times for different random splits of the data.

Other methods, including *projection pursuit* methods, clustering algorithms and *self-organizing maps* (see Chapter 6, Section 6.2.2.3), can also be used for visualization of clusters. We would like to refer to Ripley (1996), Hahne et al. (2006) and Venables and Ripley (2002) for technical details and discussion of applications.

Finally, non-linearity in the data structure has been addressed in the recent development in MDS research, e.g., *isometric mapping of data manifolds* Isomap (Tenenbaum et al., 2000) and a concurrent paper on *locally linear embedding* LLE (Roweis and Saul, 2000). The complex data sets that motivated the research included images of human faces and handwriting, which are represented by vectors of pixel intensities in high dimensions. Because of the inherent structures in the images, there exist strong correlations between images. This generates data points that lie on or close to a smooth low-dimensional manifold. The approach of Tenenbaum et al. (2000) preserves the geodesic distance (the shortest distance along the manifold) instead of transformed Euclidean distance. Alternatively, the approach of Roweis and Saul (2000) also employs the knowledge of the manifold structure of the data. Given that the manifold is well sampled, each data point and its neighbors lie approximately on a plane. LLE constructs a neighborhood preserving mapping. Both approaches share a general principle of manifold learning that the aggregate of overlapping local neighborhoods, if analyzed collectively, provides information of global geometry.

5.6 Case Study: MDS on Mass Spectrometry Data

For our case study, we will use a spike-in data set that was obtained from patients' samples to which five known proteins were added in combinations of planned different concentrations. Spike-in experiments are a means to assess accuracy and precision of a biological analytical technique. In this exercise we will use MDS to project the spectra of this data set to $\Re^3$ or $\Re^2$ to see whether distinct clusters of samples can be visualized and if so, whether the clusters correspond to the true group labels (the combinations of proteins and concentrations added). The knowledge of the true membership of samples will serve the illustrative purpose of visualization tools used.

The data set of the spike-in experiment is gathered from anonymous prostate cancer patients (Park et al., 2005). According to the paper of Park et al. (2005), plasma samples from 91 prostate cancer patients were divided into seven age matched groups, each with 13 patients. The groups were labeled A-G. Groups B-F were spiked with five proteins at 1, 2, 5, and 10 times the respective minimal concentrations using a design depicted in Table 5.1. As shown in Table 5.1, one of the proteins was left out and the remaining four were added at the different concentrations in each of the groups resulting in no identical groups. The minimal concentration of each of the spiked proteins allowing for a detectable peak in plasma had been experimentally determined and is labeled as 1. Group A was not spiked and Group G contained all five proteins at maximal (10) concentrations. Since the patients were selected to be similar in terms of disease and were matched by age across groups, the only difference among the groups is the addition of different combinations of the five proteins at certain concentrations.

Table 5.1. Design and the protein concentration, proteins 1 = Ubiquitin (1 fmol/uL), Cytochrome/Lysozyme/Myoglobin (10 fmol/uL), Trypsinogen(100 fmol/uL)

Group	Cytochrome c	Ubiquitin	Lysozyme	Myoglobin	Trypsinogen
A	0	0	0	0	0
B	0	1	2	5	10
C	1	2	5	10	0
D	2	5	10	0	1
E	5	10	0	1	2
F	10	0	1	2	5
G	10	10	10	10	10

Following the addition of proteins, 20 μL of each plasma sample was diluted with 30 μL 9 M urea and incubated at 4°C for 30 minutes in order to denature proteins. The samples were further diluted with 150 μL 1 M urea and subsequently stored at -80°C until analyzed by surface-enhanced laser desorption/ioniziation time-of-flight (SELDI-TOF) mass spectrometer, using a Biomek 2000 (Beckman Coulter, Fullerton, CA), CM10 ProteinChip Arrays (Ciphergen Biosystems, Freemont, CA). The samples were analyzed on a PBSIIc SELDI-TOF mass spectrometer (Ciphergen) according to the manufacturer's instructions at a laser setting of 190, detector setting of 7, and a digitizer rate of 1 000.

For the purpose of illustration, we will use three groups of A, D and G. Raw spectra were baseline-corrected and normalized using the *Bioconductor* R package PROcess (Li et al., 2005). The distance matrix was computed from the data matrix of processed spectra, $\mathbf{X}_{p\times n}$, where $p = 119\,400$ and $n = 39$. Classical MDS was then applied.

To visualize the results, interactive tools exist, for example, *ggobi*, that enables a data analyst to examine the data cloud in $\Re^3$ from all possible angles. For static presentation of the MDS results of the case study, we chose to use two R tools, pairs plot and the R package scatterplot3d. The R function pairs can be used to provide 2-D marginal views of MDS results for $d \geq 3$. Although human eyes are able to perceive depth, at any given angle they get a 2-D snapshot of an object. If it is a foreign object we have no experience with, we may need to "go around it" and take a series of such "snapshots" to arrive at a whole view of it. A collection of 2-D marginal views of the first few principal coordinates may be more informative than the 2-D rendering of the MDS results in understanding relationships between objects, especially when the first two directions do not explain the majority of the variation in the data. The same can be said about the 3-D rendering of an MDS result and its associated 3-D marginal views.

The R function scatterplot3d allows the viewer to rotate the data cloud in $\Re^3$ around the z-axis. By using its angle option, a data analyst is able to explore the data cloud from a set of consecutive angles for potential patterns. With minimal programming effort, one can easily rotate the data cloud around either x- or y-axis, if so desired, by passing the coordinates corresponding to x- or y-axis as the last column of the coordinates matrix.

Figure 5.2 presents the set of 2-D marginal views of the first four principal coordinates resulting from the classical MDS scaling applied to the 39 spectra.

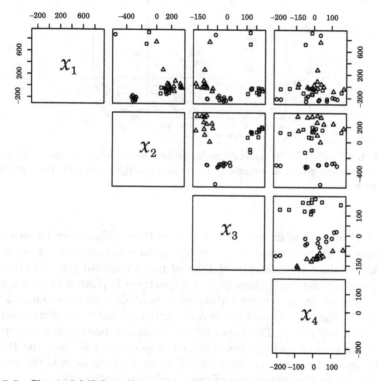

Fig. 5.2. Classical MDS scaling results of 39 spectra from groups A, D and G. Circles represent group A, squares group D and triangles group G. Each group has 13 spectra.

The first plot in Figure 5.2 is actually the 2-D MDS representation of the 39 spectra, in which we see two distinct groups, A alone, and D and G together. There are four points that appear to be outliers. Among all marginal plots, the plot of the second and the third principal coordinates shows the best separation of the three groups. However other than the outlier of group A, the other three outliers are not so obvious in this marginal view. In a 2-D representation, objects may appear to be very close but can be very far apart in the original space because the "depth" is not reflected.

Figure 5.3 shows the same classical MDS scaling results of 39 spectra in $\Re^3$. We see three clusters and four spectra that are far from their own clusters.

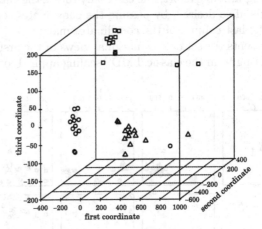

Fig. 5.3. Classical MDS scaling results of 39 spectra from groups A, D and G. Circles represent group A, squares group D and triangles group G. Each group has 13 spectra.

It is of interest to determine how each of them differs from its own group members. Spectra of each group are biological replicates (in the sense that they are obtained from samples of different individuals but yet under the same experimental treatment). Ideally, if one spectrum is plotted against another from the same group, we would expect a tight data cloud around the 45° line. The outlier spectrum of group A is plotted against three other spectra of group A in Figure 5.4. The upper right off-diagonal panels show the spectra against each other and the lower left off-diagonal panels show the Pearson correlation coefficients between pairs of spectra, with the sizes of the numbers proportional to the magnitudes of the correlation coefficients.

We observe that the outlier has unusually high intensities where the other spectra have intensities close to zero. As a result it has poor correlation with the rest of the spectra, while correlations between the other spectra are rather good. We suspect that it is due to the chemical and electronic noise at small mass-over-charge (m/z) values (Fung and Enderwick, 2002; Baggerly et al., 2004). If we use an arbitrary cutoff point of 1 000 and plot the portion of spectra with m/z values above 1 000 against each other, we see in Figure 5.5

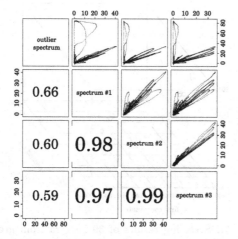

Fig. 5.4. The outlier in group A and three other spectra from the same group are plotted against each other. The lower left panels show the Pearson correlation coefficients of pairs of spectra.

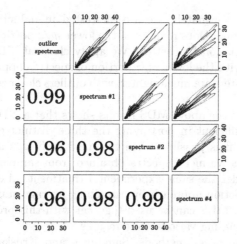

Fig. 5.5. The outlier in group A and three other spectra from the same group are plotted against each other, after ignoring the portion with m/z below 1000. The lower left panels show the Pearson correlation of pairs of spectra.

that the data cloud is more or less around the 45° line, and the correlations between the outlier and the other spectra are greatly improved.

We repeat the same exercise for the other three outliers. We find that removing the lower m/z portion of spectra helps to reduce noise in the data in the sense that truncated outlier spectra resemble the members of their

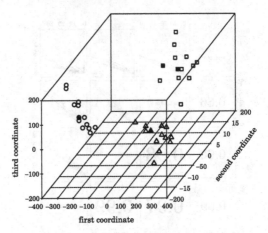

Fig. 5.6. Classical MDS scaling results of 39 spectra from groups A, D and G, ignoring the portion of spectra with m/z below 1000. Circles represent group A, squares group D and triangles group G. Solid symbols represent the outliers detected earlier. Each group has 13 spectra.

respective groups. Ultimately, we are interested in reduction of noise and enhancement of signal. Does removal of the lower m/z portion of spectra help to define object relationship better? We apply classical MDS scaling to the spectrum matrix with the lower m/z portion of spectra clipped off. Figure 5.6 shows that the clusters are much tighter, with outliers closer to their respective groups.

A `pairs` plot of the above MDS results shows that a 2-D MDS plot (Figure 5.7) is now sufficient in portraying the three distinct groups. A formal selection of a cutoff is possible when technical replicates are available. By technical replicates we mean spectra obtained from samples from the same individual and under the same experimental treatment. Li et al. (2005) proposed a cutoff selection approach by examining average standard deviations for various cutoffs. MDS can be used together with this process to examine the effects of eliminating various noise regions.

We applied two other methods, Sammon's and Kruskal and Shepard's algorithms, implemented as `sammon` and `isoMDS` of the `MASS` library, to this case study and obtained similar results.

5.7 Lessons Learned

From the case study we see that MDS, coupled with some basic graphical tools, can be employed to assess data quality and explore object relationships in high dimensions. Particularly, it can be used to examine the effect of removing certain portions of spectra. It is known that mass spectra are not reliable

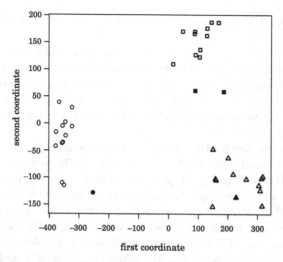

Fig. 5.7. Classical MDS scaling results of 39 spectra from groups A, D and G, ignoring the portion of spectra with m/z below 1 000. Circles represent group A, squares group D and triangles group G. Solid symbols represent the outliers detected earlier. Each group has 13 spectra.

throughout the whole m/z range for two reasons. First, when a time-of-flight (TOF) analyzer is used, the m/z values are obtained from an equation that has been established using a set of proteins or peptides (calibrants). The m/z values well outside the range of the m/z values of calibrants are extrapolations from the equation and may have large errors. This is evidenced by negative m/z values of spectra from some experiments. Secondly, chemical noise and ion overload cause the baseline of a spectrum to elevate. The chemical and electronic noise has a much larger effect on small mass-over-charge (m/z) values (Fung and Enderwick, 2002; Baggerly et al., 2004). As illustrated in our case study, it is important to identify and remove such regions so that interesting data relationships rather than artifacts can emerge through data exploration.

5.8 List of Tools and Resources

This section provides pointers to resources and tools related to the discussed and to the alternative methods. It also points to our Web site for color plots included in this chapter.

The visualization tool *ggobi* or *Rggobi* is an open source, *interactive* program for exploring high-dimensional data. It is available from `http://www.ggobi.org`. Interested readers may want to peruse the paper of Buja and Swayne (2002), for discussions of issues relating to MDS algorithms, interpretation of MDS results, local minima, indifferentiation, diagnostics of configuration and use of the software.

Our case study was conducted using R (Ihaka and Gentleman, 1996), a language and environment for statistical computing and graphics, which is available as free software under the terms of the Free Software Foundation's GNU General Public License in source code form from http://www.r-project. org. R can be considered as a dialect of S; most code written for S runs unaltered under R.

The R function cmdscale from the stats package performs classical scaling, and sammon and isoMDS of the R package MASS perform Sammon and Shepard and Kruskal scaling, respectively. The R plotting function pairs is also from the stats package. R package scatterplot3d generates scatter plot in $\Re^3$ and data points can be rotated through the angle option. Both pairs and scatterplot3d can be used to visualize an MDS analysis result. R and the aforementioned R packages are available for download from the above link. Examples for the use of the MDS functions can be found in Venables and Ripley (2002).

R based packages for handling, processing, analysis, and annotation of genomic and proteomic data are available from *Bioconductor*, http://www. bioconductor.org. *Bioconductor* is an open source and open development of software project for data from genomic and proteomic experiments. The packages genefilter and geneplotter are available from the *Bioconductor* Web site.

Although the plots in this book are produced in black and white, we recommend the use of colors in actual data analysis for better presentation and interpretation. The colored versions of the plots in this chapter can be found at http://biowww.dfci.harvard.edu/~xiaochun.

5.9 Conclusions

MDS is introduced and applied to a SELDI-TOF MS data set. In the case study, we found that MDS is useful for quality assessment of data, and presentation of object relationships. MDS can be used iteratively to assess the effects of pre-processing of spectra, in particular, cutoff selection. The low-dimensional representation of MDS results is helpful in exploration of object relationships. If the first two or three principal coordinates are insufficient in representing the data, other visualization tools need to be employed. We found pairs plot helpful in examining the data to a certain extent, because it gives marginal views of MDS results in three and higher dimensions.

Class discovery is indeed interesting, however, features, such as genes and peptides, which dictate object class membership, are of ultimate interest to biologists. Marker genes or peptides may shed light on disease genesis and evolution, serve as biomarkers for diagnosis and provide clues to targeted medicine. Although MDS is mostly used as an unsupervised data exploration tool, recent effort has been undertaken to incorporate feature selection into MDS. Simila and Tikka (2005) combined feature selection using *least angle regression* (LARS) and visualization via MDS. Their algorithm, MRSR, iterates between minimizing Sammon's fit criterion and sequential additions to the feature space. The resulting solution includes a small percentage of the original variables from a data space contributing to a projection into a two-dimensional representation of the data. Other feature selection strategies, for example cross-validation, should also be viable, to avoid overfitting.

References

Aggarwal, C.C., Hinneburg, A., and Keim, D.A. (2001). On the surprising behavior of distance matrics in high dimensional space. In *Proc. 8th Int. Conf. Database Theory (ICDT)*, pages 420–434.

Baggerly, K.A., Morris, J.S., and Coombes, K.R. (2004). Reproducibility of SELDI-TOF protein patterns in serum: comparing datasets from different experiemnts. computational challenges and solutions. *Bioinformatics*, 20:777–785.

Buja, A. and Swayne, D.F. (2002). Visualization methodology for multidimensional scaling. *J. of Classification*, 19:7–43.

Choe, G., Horvath, S., Cloughesy, T.F., Crosby, K., Seligson, D., Palotie, A., Igne, L., Smith, B.L., Sawyers, C.L., and Michel, P.S. (2003). Analysis of the phosphatidylinositol 3'-kinase signaling pathway in glioblastoma patients in vivo. *Cancer Res.*, 63:2742–2746.

Cox, T. F. and Cox, M. A. (2001). *Multidimensional Scaling*. Chapman and Hall/CRC.

Eisen, M.B., Spellman, P.T., Brownand, P.O., and Botstein, D. (1998). Cluster analysis and display of genome-wide expression patterns. *Proc. Natl. Acad. Sci. USA*, 95(25):14863–14868.

Fung, E. T. and Enderwick, C. (2002). Proteinchip clinical proteomics: Computational challenges and solutions. *Comp. Proteomics Supplement*, 32:34–41.

Gentleman, R., Carey, V. J., Huber, W., Irizarry, R. A., and Dudoit, S. (2005). *Bioinformatics and Computational Biology Solutions Using R and Bioconductor*. Springer.

Hahne, F., Huber, W., and Gentleman, R. (2006). Visualizing genomic data. Technical report, Bioconductor Project Working Papers. Working Paper 10.

Ihaka, R. and Gentleman, R. (1996). R: A language for data analysis and graphics. *J. Comp. Graph. Stat.*, 5:299–314.

Kruskal, J.B. (1964a). Multidimensional scaling by optimizing goodness-of-fit to a nonmetric hypothesis. *Psychometrika*, 29:1–27.

Kruskal, J.B. (1964b). Nonmetric multidimensional scaling: A numerical method. *Psychometrika*, 29:115–129.

Krzanowski, W. J. (2006). Sensitivity in metric scaling and analysis distance. *Biometrics*, 10:239–244.

Li, X., Gentleman, R., Lu, X., Shi, Q., Iglehart, J.D., Harris, L., and Miron, A. (2005). SELDI-TOF MS protein data. In Gentleman, R., Carey, V., Dudoit, S., Huber, W., and Irizarry, R.A., editors, *Bioinformatics and Computational Biology Solutions using R and Bioconductor*. Springer.

Mardia, K.V., Kent, J.T., and Bibby, J.M. (1979). *Multivariate Analysis*. Academic Press.

Park, Y., Downing, S. R., Li, C., Hahn, W. C., Kantoff, P. W., and L.J.Wei (2005). Simultaneous and exact interval estimates for the contrast of two groups based on an extremely high dimensional response variable: Application to mass spec data analysis. *Bioinformatics*, page In press.

Ripley, B.D. (1996). *Pattern Recognition and Neural Networks*. Cambridge University Press.

Roweis, S. and Saul, L. (2000). Nonlinear dimensionality reduction by locally linear embedding. *Science*, 290:2323–2326.

Shepard, R.N. (1962a). The analysis of proximities: Multidimensional scaling with an unknown distance function I. *Psychometrika*, 27:125–140.

Shepard, R.N. (1962b). The analysis of proximities: Multidimensional scaling with an unknown distance function II. *Psychometrika*, 27:219–246.

Simila, T. and Tikka, J. (2005). Multiresponse sparse regression with application to multidimensional scaling. In Duch, W., Kacprzyk, J., Oja, E., and Zadrozny, S., editors, *Lecture Notes in Computer Science*. Springer.

Tenenbaum, J.B., Silva, V. De, and Lanford, J.C. (2000). A global geometric framework for nonlinear dimensionality reduction. *Science*, 290:2319–2323.

Venables, W.N. and Ripley, B.D. (2002). *Modern Applied Statistics with S*. Springer.

6

Clustering – Class Discovery in the Post-Genomic Era

Joaquín Dopazo

Department of Bioinformatics, Centro de Investigación Príncipe Felipe, E46013, Valencia, Spain.
jdopazo@cipf.es

6.1 Introduction

From a historical perspective we can distinguish an initial period in the DNA microarray technology in which almost all publications were related to reproducibility and sensitivity issues. Thus, many classical microarray papers dating from the late nineties were simple proof-of-principle experiments (Eisen et al., 1998; Perou et al., 1999), in which only cluster analysis was applied in order to check whether differences at gene expression level could reproduce macroscopic observations. Later, specificity became a main concern as a natural reaction against quite liberal interpretations of microarray experiments made by some researchers, such as the fold change criterion to select differentially expressed genes. It soon became obvious that genome-scale experiments need to be carefully analyzed, because many apparent associations happened merely by chance when large amounts of data were studied (Ge et al., 2003). In this context, different methods for the adjustment of p-values, which are considered standard today, started to be extensively used (Benjamini and Yekutieli, 2001; Storey and Tibshirani, 2003). More recently, the use of microarrays for building predictive models of clinical outcomes (van't Veer et al., 2002), albeit not being free of criticisms (Simon, 2005), fueled the use of the technology because of its practical implications. There are still some concerns with the cross-platform coherence of results, but it seems clear that intra-platform reproducibility is high (Moreau et al., 2003), and, although the overlap between the lists of genes differentially expressed among platforms was low, the enrichment in biologically relevant labels emerging from these lists was consistent (Bammler et al., 2005). This fact clearly points to the importance of the interpretation of experiments in terms of their biological implications instead of restricting them to a mere comparison of lists of gene identifiers (Al-Shahrour and Dopazo, 2005; Al-Shahrour et al., 2005b).

Despite the fact that clustering is one of the most popular methodologies and the first one in being used in the field of microarray data analysis (Quackenbush, 2001; Slonim, 2002), it has often been improperly used (Simon et al., 2003). The literature on DNA microarrays provides numerous examples for the inadequate use of clustering for tackling problems of class comparison. Although cluster analysis is appropriate for class discovery, it tends to be inefficient for class comparison or class prediction. An important caveat when analyzing DNA microarray experiments is that, although these are not based on gene-specific mechanistic hypotheses, they must be designed with clear objectives. Three typical types of objectives are *class comparison, class prediction* and *class discovery* (Golub et al., 1999). *Clustering*, also known as unsupervised analysis, belongs to this last category because no previous information about the class structure of the data set is used in the study. Cluster analysis makes reference to an extensive set of methods for partitioning samples into groups on the basis of their respective differences, referred to as distances (D'Haeseleer, 2005). Usually, the distance measures are computed with regard to the complete set of genes represented on the array. Clustering can be done on the experiments (based on all the genes) or on the genes (across all the experiments). Although the methods used can be exactly the same, a note of caution must be introduced here because it is not uncommon that a given class of experiments (disease, molecular subtype, etc.) is distinguished by a relatively small number of genes, whose effect may end up being diluted by the irrelevant genes. To circumvent this problem, there exists a family of clustering methods, generically known as *biclustering*, in which the aim is to find groups of genes with coordinated expression only across a subset of experimental conditions (Cheng and Church, 2000; Lazzeroni and Owen, 2002; Tanay et al., 2002; Sheng et al., 2003).

There are other types of data that deserve particular attention: Time series or dose-response data. In this case, clustering of experiments is meaningless because there are sequential data and one is typically interested in clustering genes across all the time (or dosage) points. Recently, time series are gaining importance because the experimental methods for synchronizing cell cultures are becoming more accurate, constituting nowadays a 30% of the total number of DNA microarray experiments published (Simon et al., 2005). While typical microarray assays are designed to study static experimental conditions, in time series a temporal process is measured. Time series offer the possibility of identifying the dynamics of gene activation, which might allow to infer causal relationships. An important difference between these two types of experiments is that, while static data from a sample population (e.g., diseased cases, healthy controls, etc.) are assumed to be independent, time series data are characterized by displaying a strong autocorrelation between successive points (Bar-Joseph, 2004). Initially, time series were analyzed using methods originally developed for independent data points (Spellman et al., 1998; Zhu et al., 2000). More recently, algorithms were developed to specifically address this type of data. Different clustering methods specially designed for time series data have been recently proposed. Among these, clustering based on the dynamics of the expression patterns (Ramoni et al., 2002), clustering using a

Hidden Markov model (Schliep et al., 2003), and clustering specifically devised for short time series (Ernst et al., 2005) can be cited.

Once the clustering has been performed the following questions arise: Is the partition obtained relevant? Is there a "better" partition involving more or less clusters or a different distribution of the items within the clusters? Since most of the clustering algorithms do not include any type of measure of the reliability of the clusters obtained, these questions have to be addressed a posteriori. There are different criteria to estimate the quality of the clustering obtained (Kerr and Churchill, 2001; Azuaje, 2002; Dudoit and Fridlyand, 2003; Handl et al., 2005) and some programs (e.g., the CAAT in GEPAS (Montaner et al., 2006)) offer the possibility of obtaining cluster quality indexes. Given that some methods require that the number of clusters is predefined, (e.g., *k*-means or self-organizing maps), the exact determination of the number of clusters in the context of microarray data is a major concern, which has been specifically addressed by different authors (Horimoto and Toh, 2001; Dudoit and Fridlyand, 2002; Bolshakova and Azuaje, 2006).

But, why should we expect to find groups of co-expressed genes or a class structure in our experiments? Genes do not operate alone in the cell, but in a sophisticated network of interactions that we only recently start to decipher (Rual et al., 2005; Stelzl et al., 2005; Hallikas et al., 2006). It has been a long recognized fact that co-expressed genes tend to play some common roles in the cell (Stuart et al., 2003; Lee et al., 2004). Ultimately, it is this common functionality that we aim to understand when we face a clustering problem. Thus, an important and non-negligible last step of any clustering analysis (and, in general, of any DNA microarray experiment) is the *functional interpretation* (Al-Shahrour and Dopazo, 2005). There are a number of tools specially designed to search for significant enrichment of biological terms – usually gene ontology terms (Ashburner et al., 2000), but others can be used – in sets of genes (Khatri and Draghici, 2005). Typically, one set of genes is tested against the rest of genes in the array. This set of genes can be, more precisely, a cluster of co-expressed genes (Al-Shahrour et al., 2004), and the result produced accounts for the functional roles played by the genes in the cluster. There are different tools that allow to easily link results of clustering methods to algorithms for functional annotation, such as the GEPAS (Herrero et al., 2003, 2004; Vaquerizas et al., 2005; Montaner et al., 2006).

Recently, biological annotations (e.g., GO, KEGG pathways, etc.) have been used for cluster validation (Bolshakova et al., 2005) and, even more importantly, biological information (Huang and Pan, 2006; Pan, 2006) or phenotypic information (Jia and Xu, 2005) have been used as a constitutive part of clustering algorithms.

Clusters can be obtained in numerous different ways. There are many distinct algorithms for measuring distances among genes and many procedures for partitioning the data. In addition, most of the clustering methods do not provide any measurement of the reliability of the results obtained. This apparent diversity of ways for approaching the same problem, together with the lack of information on the reliability of the results obtained has attracted over the clustering an undeserved reputation of subjective analysis strategy. Understanding the basis of the distance metrics and the partitioning procedures and being aware of their limitations will provide the fundaments for a proper and reasonable class discovery analysis.

6.2 Basic Concepts

Despite the large number of clustering methods and the new methods proposed in the field of DNA microarray data analysis (Heyer et al., 1999; Hastie et al., 2000; Yeung et al., 2001a; de Smet et al., 2002), only a subset of them have been used with some regularity in this context. Among other merits, the reason for the popularity of many methods of microarray data analysis, and clustering is not an exception, resides in its availability in standard software packages. Among the most commonly used methods we can cite hierarchical clustering (Eisen et al., 1998), k-means (McQueen, 1967), *self-organizing maps* (SOMs) (Kohonen, 1997) or *self-organizing tree algorithm* (SOTA) (Herrero et al., 2001). Implicitly or explicitly, clustering methods depend on distances between objects. Different ways of computing distances account for different biological properties of the data. In this section I will review different distance metrics, distinct clustering algorithms, different ways of estimating cluster quality and algorithms for the functional annotation of clustering results.

6.2.1 Distance Metrics

In a widely accepted standard representation, microarray experiments are two-dimensional matrices of gene expression values in which columns correspond to genes and rows to experiments. Thus, the identification of genes with coordinated expression across the experiments or, alternatively, the identification of groups of experiments with similar expression values for all the genes is achieved through the comparison of the column or row vectors, respectively, by means of a distance function. The choice of such distance function depends on the biological property that the researcher considers. There are two types of distances extensively used in the comparison of expression profiles: *Euclidean distance* and *Pearson coefficient of correlation*.

Euclidean distance is obtained as the square root of the summation of the squares of the differences between all pairs of corresponding gene expression values (rows or columns). Euclidean distance computes the geometric distance between two points in an n-dimensional space (n being the size of the vectors – row or column – involved in the comparison). Thus, pairs of genes (or experiments) whose components display similar magnitude of expression are considered similar by this distance.

Although this property may be useful in some cases, it seems more relevant, from a biological point of view, to search for genes (or experiments), whose expression profiles display a similar overall trend, irrespective of their absolute values. The Pearson correlation coefficient (r) measures this property. It provides values between -1 (negative correlation) and 1 (positive correlation). The more the two expression profiles display the same trend, the closer to 1 is the r-value. This measure of similarity in the shapes of two profiles, while not taking the magnitude of the profiles into account, suits well the biological intuition of coexpression (Eisen et al., 1998). Euclidean distance can be used for obtaining correlations if the data are properly transformed (standardized, that is, subtracting the mean and dividing by the variance). Then the Euclidean distance between two points x and y relates to correlation as $(x - y)^2 = 2(1 - |r|)$ (Alon et al., 1999).

Most of the distances found in the microarray-related literature are derived from the Euclidean distance or from the correlation coefficient. Also some non-parametrical distances have been applied, such as the *Spearman rank correlation* (Kotlyar et al., 2002) or *jackknifed correlation coefficient* (Heyer et al., 1999). (More distance metrics can be found in Chapter 7, Table 7.2, page 157).

However, there are other different scenarios beyond the simple coexpression whose exploration is of much interest from a biological point of view. A very interesting property of the correlation coefficient is that it can be used to detect negatively correlated expression profiles. The study of such negative correlations can be very useful for identifying control processes that antagonistically regulate downstream pathways.

6.2.2 Clustering Methods

According to the final representation of the results, data can be clustered in two different ways: In a hierarchical or in a non-hierarchical manner. Hierarchical clustering allows detecting higher-order relationships between clusters of profiles whereas most of the non-hierarchical classification techniques allocate profiles into a predefined number of clusters, without any assumption on the inter-cluster relationships (see Figure 6.1a). Many authors prefer hierarchical clustering because it allows to explore the entire hierarchy of relationships at different levels. There are distinct clustering methods based on different ways of aggregating data, which use (implicitly or explicitly) different distance functions. Without the aim of producing an exhaustive enumeration of them, here I will briefly review some of the most commonly used and most relevant clustering methods. In a quick review of 1157 papers found in Pubmed using "cluster and microarray" as keywords, I have found that 74% used hierarchical clustering, 15% used k-means, 6% used SOM, 2% used SOTA, another 2% used model-based clustering, and in the remaining cases other alternative methods were used. Although these figures can change depending on the keywords used for finding the papers, they give an approximate idea on the relative actual usage of each procedure.

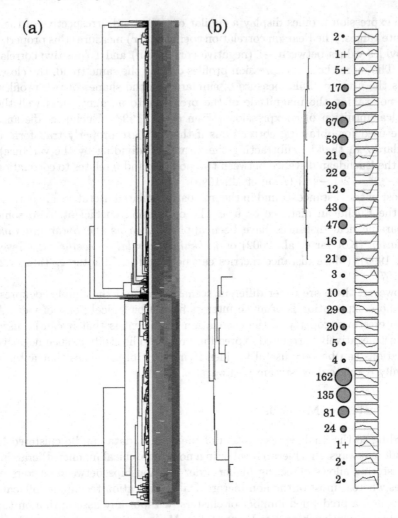

Fig. 6.1. Different clustering methods applied to cluster genes. a) Aggregative hierarchical clustering, and b) SOTA with default parameters.

6.2.2.1 Aggregative Hierarchical Clustering

Aggregative hierarchical clustering (Eisen et al., 1998) is one of the preferred choices for the analysis of patterns of gene expression (Quackenbush, 2001; D'Haeseleer, 2005). Standard aggregative hierarchical clustering produces a representation of the data with the topology of a binary tree, in which the most similar patterns are clustered in a hierarchy of nested subsets (Sneath and Sokal, 1973). Figure 6.1a shows a typical output of produced by the method. In aggregative hierarchical clustering, each vector (gene or experi-

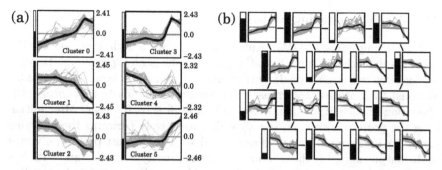

Fig. 6.2. Different clustering methods applied to cluster genes. a) k-means with $k = 6$, and (b) SOM with a 4×4 output map with hexagonal neighborhood.

ment) is initially assigned to a single cluster; at each step, the distance between every pair of clusters is calculated and the pair of clusters with the smallest distance is merged; the procedure is iteratively carried on until all the data are grouped into a single cluster. Depending on the way in which vectors are merged into a cluster and the distance of the new cluster to the rest of items (also known as linkage distance) is calculated, different variants of the method can be distinguished. This linkage distance can be calculated as the shortest distance of any of the two joined members (*single linkage*), the largest distance (*complete linkage*) or either weighted or unweighted averages (*average linkage*).

After the full tree is obtained, the determination of the final partition is achieved by "cutting" the tree at a certain level or height, which is equivalent to putting a threshold on the pairwise distance between clusters. Note that the decision of the final partition is thus rather arbitrary.

6.2.2.2 k-Means

The k-means algorithm (McQueen, 1967) requires the specification of the number of clusters, k, into which the objects are going to be partitioned. Then the mean vector for each of the k clusters, the *seed*, is initialized either by direct assignment (e.g., from the input) or by random generation (random initial seeds). Then, the algorithm proceeds through an iterative procedure, consisting of the following two steps: (1) Using the given mean vectors, the algorithm assigns each gene (or experiment) to the cluster with the closest mean vector. (2) The algorithm recalculates the mean vectors (which are the sample means) for all the clusters. The iterative procedure ends when all the mean vectors of the clusters remain constant or do not change significantly. Figure 6.2a shows the output of the algorithm in a data set using $k = 6$.

6.2.2.3 Self-Organizing Maps

Self-organizing maps (SOM) (Kohonen, 1997) are a technique to visualize the high-dimensional input data (in our case, the gene expression data) onto an (usually two-dimensional) output map of prototype vectors (also called neurons) by a process known as *self-organization*. Similarly to k-means, the dimension of the output map needs to be specified by the user. After initializing the prototype vectors, the algorithm iteratively performs the following steps. (1) Every input vector is associated with the closest prototype vector of the output map, (2) the components of the prototype vector (and with less intensity the prototype vectors in the neighborhood) are updated according to a weighted sum of all the input vectors that are assigned to it. This process is repeated until the prototype vectors of the output node converged to a constant value. During the clustering process the prototype vectors are pulled towards the regions of the space that are more densely populated by the input vectors. Figure 6.2b shows a typical output of SOM using an output map of 4×4 with an hexagonal neighborhood.

6.2.2.4 Self-Organizing Tree Algorithm

The *self-organizing tree algorithm* (SOTA) (Dopazo and Carazo, 1997; Herrero et al., 2001) is a different type of self-organizing neural network based on the SOM, but implementing a binary tree topology, instead of the classical two-dimensional grid, and a different strategy of training. The iterative procedure, with the application of the self-organization principle to the prototype vectors, is similar to the case of SOM. The differences reside in the fact that the unique prototype vectors directly updated are the leaves of the tree structure. The neighborhood is defined through the tree topology. After convergence of the network, the prototype vector containing the most variable population of expression profiles (variation is defined here by the maximal distance between two profiles that are associated with the same prototype vector) is split into two sister vectors (causing the binary tree to grow), hereafter the entire process is restarted. The algorithm stops (i.e., the tree stops growing) when a threshold of variability is reached for each prototype vector. Hence, the number of clusters does not need to be specified in advance. The determination of the threshold of variability involves the actual construction of a randomized data set. In contrast to hierarchical clustering, which is an aggregative method, the SOTA is divisive. Figure 6.1b shows an example of clustering of genes obtained with SOTA.

6.2.2.5 Model-Based Clustering

Although *model-based clustering* has already been used in the past in other fields, its application to microarray data is relatively recent (Yeung et al., 2001a; Ghosh and Chinnaiyan, 2002; McLachlan et al., 2002). In contrast to the clustering methods described so far, model-based methods provide a consistent statistical framework for obtaining data partitions. The basic assumption in model-based clustering is that the data are generated by a mixture of a finite number of underlying probability distributions, where each distribution represents one cluster. Model-based clustering methods face the problem of associating every gene (or experiment) with the best underlying distribution in the mixture, and at the same time, finding out the parameters for each of these distributions. Different approximations can be used to infer these parameters. Gaussian mixture models have been applied with success to microarray data clustering (Yeung et al., 2001a). On the other hand, problems such as the estimation of the number of clusters can be solved in a more efficient way using a Bayesian framework (Vogl et al., 2005).

6.2.3 Biclustering

As previously mentioned, biclustering methods search for genes with coordinated expression across a subset of experiments. While in the beginning clustering algorithms were applied to both genes and experiments of the microarray matrix to reorganize data and thus visualize patterns common to genes and experiments (Alon et al., 1999; Getz et al., 2000), soon algorithms specifically designed for biclustering, such as the *Samba*, methods based on graph theory (Tanay et al., 2002), the *iterative signature algorithm* (ISA) (Ihmels et al., 2002) or mixtures of normal distributions (Lazzeroni and Owen, 2002) were proposed. Recently, model-based algorithms providing a more rigorous statistical framework have been proposed (Barash and Friedman, 2002; Sheng et al., 2003).

6.2.4 Validation Methods

As previously mentioned, validation of the relevance of the cluster results is of paramount importance given that most clustering methods do not provide any clue on reliability. Validation can be based on either external or internal criteria. In the first case some gold standard is chosen and its agreement with the partition obtained by the clustering method is taken as a support for such a partition. Usually, biological information (gene ontology, pathways, etc.) is used for this purpose (Tavazoie et al., 1999; Toronen, 2004; Al-Shahrour and Dopazo, 2005). Internal criteria for statistical cluster validation imply the assessment of cluster coherence using different measures that compare inter- to intra-cluster variability, such as *silhouette coefficient* (Rousseeuw, 1987), *Dunn-like indices* (Azuaje, 2002), *connectedness* or *separation measures* (Handl et al., 2005). Another internal criterion consists of testing the stability or the robustness of a cluster result when noise is deliberately added to the data (Kerr and Churchill, 2001; Dudoit and Fridlyand, 2002).

6.2.5 Functional Annotation

Clustering of microarray data produces a collection of objects (genes or experiments) based on the comparison of their expression profiles but gives no information on the functional basis for this grouping. While not much effort has been developed on the way of understanding the molecular functional basis of clustering of experiments, there are however numerous papers dealing with the issue in the case of clustering of genes. Ending up with a mere list of genes of interest is only half-way to the result of a microarray experiment. Apart from the utility that functional annotation can have as an external criterion for cluster quality, it constitutes itself an unavoidable final step of any microarray analysis. The proper interpretation of cluster analysis of microarray experiments is usually performed in two steps: In a first step, clusters of genes of interest are selected, and then the enrichment of any type of biologically relevant annotation for these genes is compared to the corresponding distribution of this annotation in the background (typically, the rest of genes). It is important to note that this comparison to the background is essential because sometimes apparent high enrichment in a given annotation is nothing but a reflect of a high proportion of this particular term in the whole genome and, consequently, has nothing to do with the set of genes of interest. There are different available tools, such as FatiGO (Al-Shahrour et al., 2004) and others (Khatri and Draghici, 2005), that estimate significant enrichment in different functionally relevant annotation terms such as GO (Ashburner et al., 2000), KEGG pathways (Kanehisa et al., 2004), etc.

6.3 Advantages and Disadvantages

The methods and algorithms previously described have been developed for situations and under assumptions that are not always fulfilled by DNA microarray data. In this section I will comment some of the positive and negative features of the methods in the light of some of the most common problems in clustering.

- **Finding the proper number of clusters.** In general, clustering methods do not define the proper number of clusters by themselves. k-means and SOM need the pre-specification of the number of clusters. Different strategies are used to circumvent this problem, but commonly different runs of the program with different values of k (in k-means) need to be evaluated with a quality cluster index to decide about the optimal number of clusters. Nevertheless, this strategy is finally computationally expensive. A similar problem affects some model-based procedures. In this case the algorithm has to compare multiple log maximum likelihood values to optimize the complexity of the model (Yeung et al., 2001a), or resampling the data set (Yeung et al., 2001b). Both strategies are very time-consuming.

On the other hand, model-based methods based on a Bayesian approach can estimate the partition with the proper number of clusters (although also at the expense of high run times). Besides, the SOTA method (Herrero et al., 2001) implements a quick permutation-based strategy that produces the partition at which the clusters contain elements with significant intra-cluster distances (that is, distances that cannot be found in random clusters).

- **Reliability of the clusters obtained.** As mentioned above, the reliability of clustering methods can be checked in different ways, based on external information or on internal properties of the partition obtained. There are several benchmarking studies that compare the relative efficiencies of different clustering methods in defining partitions in both artificial data sets and in well-known real data sets (Gibbons and Roth, 2002; Datta and Datta, 2003; D'Haeseleer, 2005; Handl et al., 2005). As general conclusion, hierarchical clustering with single linkage would not be a good choice, because of its poor performance (Gibbons and Roth, 2002; D'Haeseleer, 2005). Depending on the study, hierarchical clustering with complete or average linkage results in different performances: Sometimes one of the linkage strategies seems to work better than the alternative and sometimes not. In general, k-means, SOM and SOTA seem to exhibit a better performance than hierarchical clustering (Gibbons and Roth, 2002; D'Haeseleer, 2005; Handl et al., 2005) according to different indexes such as silhouette, Dunn, etc. It is important to note here that the performances reported for k-means and SOM refer to an unrealistic situation in which the number of clusters is provided to the method. This information is currently unknown in real scenarios. Unfortunately, there are no benchmarking studies that include model-based clustering methods to date, and only a few performance comparisons are available. Thus, for example, model-based Bayesian methods seem to perform better than k-means, even in situations in which the number of clusters (k) was provided to the method (Vogl et al., 2005).
- **Reliability of biclustering.** An interesting, although not exhaustive, comparative study has recently been published (Prelic et al., 2006). Here, the Samba (Tanay et al., 2002) and ISA (Ihmels et al., 2002) methods seem to work reasonably well in the absence of noise, and a method proposed by the authors, *Bimax*, seems to outperform them in noisy situations.
- **Run times.** Despite the advantages of model-based clustering methods (they can estimate the reliability of the partition and some versions can estimate the number of clusters and impute missing values), their extremely long run times (hours to days) and usually its requirement of powerful computers for running, represent a limitation to its application to real situations. As a general rule, methods that use pair-wise distance matrices (e.g., hierarchical clustering or k-means) have run times that are, at least, quadratic on the number of items, while methods based on the distances of the items to a number of clusters (e.g., SOM or SOTA) have almost linear

run times. Nevertheless, a data set with a number of features ranging from 20 000 to 40 000 and a number of experiments ranging from 20 to 100 can be a matter of seconds for SOM and SOTA, a few minutes for hierarchical clustering and no more than 15 minutes for k-means.

- **Interpretation of the results.** As mentioned above, the functional annotation of the partition obtained can be used as an external criterion to check the quality of the clustering obtained, but, at the same time it is crucial for obtaining a proper annotation of the results obtained. Functional annotation of clusters implies searching for enrichment of some functional terms (typically GO, KEGG pathways, etc.) in them. One important consideration in this step is the correction for multiple testing. For example, there are around 14 000 GO terms; the possibility of finding apparent enrichments in a few GO terms just by chance is high. To avoid obtaining a considerable number of false positive enrichments different methods for multiple testing adjustments can be used. Beyond the classical Bonferroni or Holm's corrections, which are extremely conservative, one of the most popular choices are the *false discovery rate* (FDR), which in addition accounts for dependencies between the data (Benjamini and Hochberg, 1995; Benjamini and Yekutieli, 2001). One of the first programs to incorporate this correction was *FatiGO* (Al-Shahrour et al., 2004), although now it is included in a number of systems (Onto-Express, GOStat, GOToolBox, Gosurfer, etc.) Despite the importance of applying such corrections, there are still programs, such as *GoMiner, DAVID, eGOn, GOTM* or *CLENCH* that do not include it yet (Khatri and Draghici, 2005).

6.4 Caveats and Pitfalls

It is worth noting that many clustering methods produce partitions even with random data. This is commonly known as the "garbage-in-garbage-out" effect in programming and points to the necessity of having some criteria in the application of these methods. There are two potential weak points in any clustering analysis: The distance function used and the algorithm for producing the partition. The combined effect of both choices (sometimes restricted by the clustering method) and the properties of the particular data set at hand will make one of the methods more efficient compared to the alternatives. Benchmarking studies, albeit not perfect, give an idea of the relative performance of the different methods under different conditions, especially when some of the conclusions are consistent across different, independent studies. In the previous section some considerations have been made on the different methods and a common conclusion was the poor performance of hierarchical clustering when single linkage was used. While hierarchical clustering with average or complete linkage seems to work well, SOM, SOTA and k-means seem to be superior according to internal indexes (Silhouette, Dunn, and other) or external criteria (enrichment of functional terms). Model-based methods (in particular Bayesian approaches) seem to show a superior performance, although run times are still excessive as to be considered feasible alternatives on many computers. Beyond the advantages and disadvantages commented in the previous section some considerations follow that deserve to be made.

6.4.1 On Distances

Usually, the distance metrics are computed with regard to the complete set of genes represented on the array. Clustering can be done on the experiments (based on all the genes) or on the genes (across all the experiments). It is not uncommon that a given class of experiments (diseases, molecular subtypes, etc.) is distinguished by a relatively low number of genes, whose effect may end up diluted among the contributions of the rest of genes. This can lead to the construction of groups based on irrelevant features unrelated to the aim of the study. And this effect represents an even greater problem when working with systems that cannot be under a strict experimental control, i.e., patients or samples directly collected from nature. In a classical paper, only two types of diffuse large B-cell lymphoma could clearly be defined while some subtypes were merged together in clusters not reflecting the clinical subtype composition of the disease (Alizadeh et al., 2000). The only way described so far to overcome this problem is via a biclustering approach. Similarly, typical distances used in microarray assume that all the vector components used in the computation are independent and this assumption clearly does not hold in the case of time series, where all the experiments are autocorrelated. In this case clustering methods specifically designed for time series should be used (Ramoni et al., 2002; Schliep et al., 2003). Moreover, microarray time series are short in comparison with typical time series in other disciplines (about 80% of microarray time series experiments involve only three to eight time points (Ernst and Bar-Joseph, 2006)), so clustering methods specifically developed for this purpose should be used (Ernst et al., 2005).

6.4.2 On Clustering Methods

A significant problem associated with k-means or SOM algorithms is the arbitrary choice of the number of clusters, since this information is commonly not available in a real class-discovery problem. In practice, this makes it necessary to use a trial-and-error approach where a comparison and validation of several runs of the algorithm with different parameter settings are necessary. A similar problem affects some versions of model-based methods, such as Gaussian mixture models (Yeung et al., 2001a), but the strategies for finding the number of clusters here (Yeung et al., 2001a,b) are enormously time-consuming.

Another parameter that will influence the result of k-means clustering is the choice of the seeds. The algorithm is hampered by the problem of local minima. This means that with different seeds, the algorithm can yield different result. This problem also applies to SOM although to a lesser extent.

Another inherent problem in SOM is that the training of the network (and, consequently, the definition of clusters) depends on the number of items assigned to each cluster. If irrelevant data (e.g., invariant, "flat" profiles) or some particular type of profile is over-represented in the data, SOM will produce an output in which this type of data will populate the vast majority of clusters. As a consequence, the most interesting profiles may appear in a few clusters and the resolution obtained for them is poorer.

In contrast to SOM, the number of nodes does not need to be initialized in SOTA. The partition obtained with SOTA is proportional to the heterogeneity of the data, but not to the number of items in each cluster. Thus, SOTA is quite insensitive to perturbing effects of big clusters on the global cluster structure and can simultaneously resolve small and big clusters. Since SOTA is a divisive method, a test can easily be coupled to the growing tree process to decide at which point the growing of the tree should be stopped because all the significant clusters have been found (Herrero et al., 2001).

6.5 Alternatives

Clustering has been extensively used over many years for different purposes and consequently many clustering methods are available (Sneath and Sokal, 1973), so an exhaustive description of alternatives falls beyond the scope of this chapter. In this chapter the clustering methods most commonly used in the field of microarray data analysis have been described. Nevertheless, other proposals have been made that, despite their potential, have not been extensively used yet.

Early from an historical perspective in microarray data analysis, the QT-Clust method was introduced (Heyer et al., 1999). This method considers each expression profile in the data and determines how many of them are within the distance specified as quality guarantee. The candidate cluster with the largest number of expression profiles is selected as the output of the algorithm. Then, the expression profiles of the selected cluster are removed, and the whole procedure starts again to find the next cluster. The algorithm stops when the number of profiles in the largest remaining cluster falls below a pre-specified threshold.

Adaptive quality-based clustering (de Smet et al., 2002) uses a heuristic two-step approach to find one cluster at a time. In the first step, a quality-based approach is performed to locate a cluster center in the area where the density (i.e., the number) of gene expression profiles has a local maximum. In the second step, the algorithm re-estimates the quality (i.e., the radius) of the cluster so that the genes belonging to the cluster are, in a statistical sense, significantly co-expressed. The cluster found is subsequently removed from the data and the whole procedure is restarted. Only clusters whose size exceeds a predefined number are reported in the output.

Contrarily to aggregative hierarchical clustering, the divisive version of this method provides a picture of the tree from lower to higher resolution, as the construction of the tree proceeds. Apart from SOTA (Herrero et al., 2001), other divisive hierarchical methods, e.g., based on the maximum entropy principle (Alon et al., 1999), have been proposed. The algorithm tries to find the most likely partition of data into sets and subsets, creating in this way a binary tree structure.

Fuzzy versions of some clustering methods have also been applied to microarray data analysis (Dembele and Kastner, 2003). The rationale behind the proposal of the use of fuzzy methods is the difficulty of defining cluster boundaries (Spellman et al., 1998). Fuzzy membership of genes should then be considered more an operative procedure than a reality. Difficulties in the placement of a gene in a cluster are due to noise and multi-functionality and can be best addressed through biclustering methods.

Furthermore, other types of distances can be mentioned. There are distances that can deal with data sets containing large numbers of measures that have a high degree of internal correlations. Correlations between experiments or genes tend to produce elliptical clusters, which cause problems to methods whose optimal performance occurs with compact, spherical clusters, such as k-means. Distances that take into account covariance between experiments, like the Mahalanobis distance (Mahalanobis, 1936), may be useful for data sets with high internal correlation. The problems that originate from the complex joint distribution of gene expression values, particularly their structure of internal correlations and non-normality, have been addressed by other researchers (Hunter et al., 2001), who argue that simple similarity metrics such as Euclidean distance or correlation similarity are suboptimal in microarray data sets and propose the use of Bayesian approaches.

6.6 Case Study

Understanding the molecular roles played by potentially relevant genes in a given experiment is still one of the most interesting objectives in many microarray experiments. One of the most popular hypotheses in microarray data analysis is that coexpression of genes across a series of experiments is most probably explained through some common functional role (Eisen et al., 1998). Actually, this causal relationship has been used to predict gene function from patterns of co-expression (Stuart et al., 2003; Lee et al., 2004).

In this case study I use data from a genome-wide study to search for the factors responsible for the transcription of the cluster of co-ordinately expressed ribosomal proteins of *Saccharomyces cerevisiae* (Rudra et al., 2005). This data set is publicly available at http://gepas.bioinfo.cipf.es/cgi-bin/ datasets. There is a step that must be taken prior to any sort of microarray data analysis: *Normalization* of the data. This step is beyond the scope of this

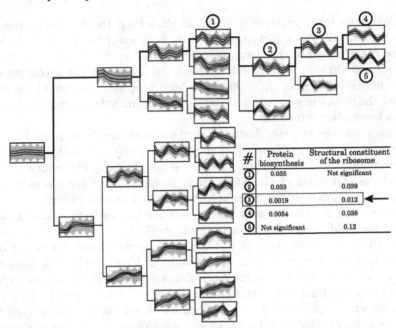

#	Protein biosynthesis	Structural constituent of the ribosome
①	0.035	Not significant
②	0.033	0.039
③	0.0019	0.012
④	0.0054	0.038
⑤	Not significant	0.12

Fig. 6.3. Clustering of gene expression profiles obtained with the SOTA method (setting the variability threshold to 80%) and represented using the CAAT tool (Montaner et al., 2006). The summarized description of the tree is obtained with the CAAT tool, with the representation of the gene expression profiles (individual gene profiles in grey and average profile in black) assigned to clusters and sub-clusters. The upper branch is developed until no more partitions are produced by the SOTA algorithm. Note how the confidence intervals for the average gene expression profile become narrower as we move towards the terminal nodes. The arrow marks the level at which the enrichment in the biological terms studied has a maximum significance.

chapter and aims to remove all the variability due to experimental manipulation and unrelated to the actual experiment. (See Chapter 3 for details on normalizing microarray data.) It can be carried out by using standard programs such as the DNMAD (Vaquerizas et al., 2004), SNOMAD (Colantuoni et al., 2002) or other programs. With the goal of finding groups of genes that co-express across the experiments, gene expression patterns were clustered using the SOTA algorithm (Herrero et al., 2001) as implemented in the GEPAS (http://www.gepas.org) suite of Web tools (Herrero et al., 2003, 2004; Vaquerizas et al., 2004; Montaner et al., 2006). Figure 6.3 shows a general view of the SOTA hierarchical tree obtained, where the top branch is shown in detail, and the terminal nodes (clusters as defined by SOTA) are at the right end. The CAAT tool allows selecting clusters and automatically submitting them to the *FatiGOplus* tool (Al-Shahrour et al., 2004, 2006) for functional analysis. When the upper terminal node is chosen, we found the GO terms "Protein biosynthesis" (FDR-adjusted $p = 0.0054$) and "Structural constituent of the ribosome" (FDR-adjusted $p = 0.0038$) significantly over-represented in the group of co-expressing genes contained in the cluster, when compared to the rest of the tree. This operation can be repeated for all the nodes of the tree and most of them will display significant over-representation of GO terms. And, what is even more interesting, we can examine internal nodes. If the internal nodes are sequentially analyzed along the branch of a tree for the enrichment in biologically relevant terms it is possible to find a level in the tree in which this enrichment is maximum (and significant). In Figure 6.3, this level in the tree that maximizes the proportion of genes annotated as "protein biosynthesis" and "structural constituent of the ribosome" is marked by an arrow. Actually, it is the parent of the level at which SOTA decides to stop growing. As we move from higher levels to lower levels of the hierarchy we find clusters with tighter co-expression, which are more likely involved in a common function. At this point, clustering based on the distance measure has a natural, functional meaning. Beyond this point, new partitions will not reflect a functional (biologically relevant) co-expression (see the two last clusters in which the p-value increases or, in some cases is non significant). Functional annotation can be considered an external cluster quality measure.

6.7 Lessons Learned

The first and most important lesson is that clustering is for class discovery (unsupervised analysis), but not for class discrimination or class prediction (supervised analysis). Although this may sound obvious, there is still an extensive misuse of these techniques (Simon et al., 2003). Clustering of genes and experiments can be carried out using exactly the same methods (applied to columns or to rows, respectively). The final partition obtained is based on equal contributions of each experiment (when clustering genes) or each gene (when clustering experiments). It is worth remembering that many genes will only introduce noise (because they represent physiological conditions or any particularity of the sample, unrelated to the biological trait we have in mind) and consequently, the partition obtained could be irrelevant from a biological point of view. Not all the experiments are equivalent in terms of their analysis. In addition to the inherent noise there are situations, such as time series or dose-response experiments, in which the data display a high internal correlation. These cases should be clustered with methods specifically designed for them (Bar-Joseph, 2004). Finding a partition requires the correct estimation of the number of clusters and its reliability. Only a few methods include these features . Among them I can cite SOTA (Herrero et al., 2001) or recent versions of model-based clustering (Vogl et al., 2005), although the latter is too demanding in computational resources to constitute an alternative. Most clustering methods require external strategies to find the optimal number of clusters and their reliability. This is nothing that should prevent one from using a particular clustering method but must be taken into account. Just trying with several k values for k-means and choosing the partition which "looks nicer" can be a interesting exploratory exercise, but is definitively not a proper way of obtaining a partition. Contrarily, irrespective of the final decision on the clustering method, SOTA, given its reliability (according internal indexes and external criteria, see Handl et al. (2005)) and speed, constitutes a good choice for a first exploration of the data. With respect to the performance of the different methods, recent comparative studies (Gibbons and Roth, 2002; Datta and Datta, 2003; D'Haeseleer, 2005; Handl et al., 2005) suggest that hierarchical clustering (with complete or average linkage), SOM and k-means (if the number of clusters is known) and SOTA tend to produce accurate partitions according to several cluster quality indexes. And last but not least, clusters of co-expressing genes represent biological processed cooperatively carried out by the genes. A proper understanding of these processes require of the application of methods that examine the biological roles jointly carried out by the genes, that is, the functional annotation of the experiment (Al-Shahrour and Dopazo, 2005; Khatri and Draghici, 2005).

6.8 List of Tools and Resources

There are different tools available and several repositories containing tools for the analysis of microarray data. The list below does not intend to be an exhaustive catalogue of these resources but contains some of the most complete and stable ones.

6.8.1 General Resources

- http://www.nslij-genetics.org/microarray/soft.html.
- http://ihome.cuhk.edu.hk/.
- http://bioinformatics.ubc.ca/resources/.

6.8.1.1 Multiple Purpose Tools (Including Clustering)

- GEPAS: A Web-based resource for microarray gene expression data analysis (Herrero et al., 2003, 2004; Vaquerizas et al., 2005; Montaner et al., 2006), which beyond clustering offers many more tools (normalization, gene selection, predictors, functional annotation, Array-CGH, etc.). http://www.gepas.org.
- INCLUSive: A Web portal for clustering and regulatory sequence analysis (Coessens et al., 2002). http://www.esat.kuleuven.ac.be/inclusive.
- Expression Profiler is a Web-based platform for microarray data analysis developed at the EBI (Kapushesky et al., 2004). http://www.ebi.ac.uk/expressionprofiler.

6.8.2 Clustering Tools

- http://homes.esat.kuleuven.be/~thijs/Work/Clustering.html: Adaptive Quality-Based Clustering (de Smet et al., 2002).
- http://www.ii.uib.no/~bjarted/jexpress/index.html: J-EXPRESS: University of Bergen, Norway.
- http://rana.lbl.gov/EisenSoftware.htm: CLUSTER, TREEVIEW: Eisen's lab at Lawrence Berkeley National Laboratory.
- http://www.genome.wi.mit.edu/MPR/software.html: GENE-CLUSTER: Whitehead Institute.
- http://gepas.bioinfo.cipf.es/cgi-bin/sotarray: SOTA (Herrero et al., 2001): CIPF, Spain. Also included in GEPAS.

6.8.3 Biclustering Tools

There are not many biclustering tools available yet. The coupled two-way analysis (Getz et al., 2000) is a simple method available at http://ctwc. weizmann.ac.il/. Also, GEMS (Wu and Kasif, 2005) is a nice example of a Web-based tool for biclustering (available at http://genomics10.bu.edu/ terrence/gems/).

6.8.4 Time Series

Time series (and dose-response) experiments are characterized by displaying a strong autocorrelation between successive points (Bar-Joseph, 2004) and must, consequently, be analyzed with algorithms that specifically take into account this fact. The algorithm STEM has been, in addition, designed for short time series and can be found at http://www.cs.cmu.edu/~jernst/ stem.

6.8.5 Public-Domain Statistical Packages and Other Tools

Probably, the most popular resource for microarray data analysis is *Bioconductor* (Gentleman et al., 2004). It is written in the popular R statistical programming language and offers many modules for the analysis of microarray data. It is available at http://www.bioconductor.org. The BRB tools, developed by the Richard Simon and Amy Peng Lam group, offer a variety of useful algorithms. Available at: http://linus.nci.nih.gov/BRB-ArrayTools. html. Additionally, there are packages in Java, which are very popular, as is the case of MEV (http://www.tigr.org/software/microarray.shtml) (Saeed et al., 2003). Java packages provide an interactive and convenient interface and can run on multiple platforms, constituting an interesting alternative to Web-based tools, which cannot offer the same degree of interactivity. The only limitation comes from the characteristics of the local computer in which the program is installed (which can be an obstacle in a non-negligible number of cases).

6.8.6 Functional Analysis Tools

- *Babelomics* (Al-Shahrour et al., 2005b, 2006) is a suite of Web tools for the functional annotation and analysis of groups of genes in high throughput experiments. Tools include: *FatiGO* (Al-Shahrour et al., 2004), *FatiGO-plus*, *Fatiscan* (Al-Shahrour et al., 2005a), *Gene Set Enrichment Analysis* (GSEA) (Subramanian et al., 2005), *Marmite*, and the *Tissues Mining Tool* (TMT). http://www.babelomics.org.
- *goCluster* simultaneously implements annotation information, clustering algorithms and visualization tools for microarray data analysis (Wrobel et al., 2005). Available at: http://www.bioconductor.org; http://www.bioz.unibas.ch/gocluster.

6.9 Conclusions

Clustering is essential for finding either (functionally related) co-expressed genes or subtypes of experiments based on their gene expression profiles. Although clustering of genes and experiments can be carried out using exactly the same methods, the final result obtained is based on equal contributions of each data component. Thus, it is worth noting that in the case of clustering of experiments many genes will only introduce noise and consequently the resulting partition can be meaningless from a biological point of view. In addition to noise, some experiments are conceptually different. Time series or dose-response experiments, for example, are characterized by the existence of a high internal correlation between consecutive experiments. These experiments must be clustered with methods specifically designed for them (Bar-Joseph, 2004). Regarding the comparative performances of the methods, hierarchical clustering (except in the case of single linkage), SOM and k-means (provided the number of clusters is known) and SOTA seem to produce reliable partitions (Gibbons and Roth, 2002; Datta and Datta, 2003; D'Haeseleer, 2005; Handl et al., 2005). Finally, methods that examine the enrichment in biologically relevant terms (Al-Shahrour and Dopazo, 2005; Khatri and Draghici, 2005) are necessary for a proper understanding of the biological processes cooperatively carried out by the genes present in co-expression clusters .

Acknowledgements

This work is supported by grants from MEC BIO2005-01078, NRC Canada-SEPOCT Spain and Fundación Genoma España.

References

Al-Shahrour, F., Diaz-Uriarte, R., and Dopazo, J. (2004). FatiGO: A web tool for finding significant associations of gene ontology terms with groups of genes. *Bioinformatics*, 20(4):578–580.

Al-Shahrour, F., Diaz-Uriarte, R., and Dopazo, J. (2005a). Discovering molecular functions significantly related to phenotypes by combining gene expression data and biological information. *Bioinformatics*, 21(13):2988–2993.

Al-Shahrour, F. and Dopazo, J. (2005). Ontologies and functional genomics. In Azuaje, F. and Dopazo, J., editors, *Data analysis and visualization in genomics and proteomics*, pages 99–112. Wiley, West Sussex, UK.

Al-Shahrour, F., Minguez, P., Tarraga, J., Montaner, D., Alloza, E., Vaquerizas, J.M., Conde, L., Blaschke, C., Vera, J., and Dopazo, J. (2006). BABELOMICS: A systems biology perspective in the functional annotation of genome-scale experiments. *Nucleic Acids Res., in press*.

Al-Shahrour, F., Minguez, P., Vaquerizas, J.M., Conde, L., and Dopazo, J. (2005b). BABELOMICS: A suite of web tools for functional annotation and analysis of groups of genes in high-throughput experiments. *Nucleic Acids Res.*, 33:W460–464.

Alizadeh, A.A., Eisen, M.B., and Davis, R.E., et al. (2000). Distinct types of diffuse large B-cell lymphoma identified by gene expression profiling. *Nature*, 403(503):511.

Alon, U., Barkai, N., Notterman, D.A., Gish, K., Ybarra, S., Mack, D., and Levine, A.J. (1999). Broad patterns of gene expression revealed by clustering analysis of tumor and normal colon tissues probed by oligonucleotide arrays. *Proc. Natl. Acad. Sci. USA*, 96(12):6745–6750.

Ashburner, M., Ball, C.A., and Blake, J.A., et al. (2000). Gene ontology: Tool for the unification of biology. *Nat. Genet.*, 25:25–29.

Azuaje, F. (2002). A cluster validity framework for genome expression data. *Bioinformatics*, 18(2):319–320.

Bammler, T., Beyer, R.P., and Bhattacharya, S. (2005). Standardizing global gene expression analysis between laboratories and across platforms. *Nat. Methods*, 2:351–356.

Bar-Joseph, Z. (2004). Analyzing time series gene expression data. *Bioinformatics*, 20(16):2493–2503.

Barash, Y. and Friedman, N. (2002). Context-specific bayesian clustering for gene expression data. *J. Comp. Biol.*, 9:169–191.

Benjamini, Y. and Hochberg, Y. (1995). Controlling the false discovery rate: A practical and powerful approach to multiple testing. *J. Roy. Stat. Soc.*, B57:289–300.

Benjamini, Y. and Yekutieli, D. (2001). The control of false discovery rate in multiple testing under dependency. *Ann. Stat.*, 29:153–157.

Bolshakova, N. and Azuaje, F. (2006). Estimating the number of clusters in DNA microarray data. *Methods Inf. Med.*, 45:153–157.

Bolshakova, N., Azuaje, F., and Cunningham, P. (2005). A knowledge-driven approach to cluster validity assessment. *Bioinformatics*, 21(10):2546–2547.

Cheng, Y. and Church, G.M. (2000). Biclustering of expression data. *Proc. Int. Conf. Intell. Syst. Mol. Biol.*, 8:93–103.

Coessens, B., Thijs, G., Aerts, S., Marchal, K., de Smet, F., Engelen, K., Glenisson, P., Moreau, Y., Mathys, J., and de Moor, B. (2002). INCLUSive: A web portal and service registry for microarray and regulatory sequence analysis. *Nucleic Acids Res.*, 31:3468–3470.

Colantuoni, C., Henry, G., Zeger, S., and Pevsner, J. (2002). SNOMAD (Standardization and NOrmalization of MicroArray Data): web-accessible gene expression data analysis. *Bioinformatics*, 18(11):1540–1541.

Datta, S. and Datta, S. (2003). Comparisons and validation of statistical clustering techniques for microarray gene expression data. *Bioinformatics*, 19(4):459–646.

de Smet, F., Mathys, J., Marchal, K., Thijs, G., de Moor, B., and Moreau, Y. (2002). Adaptive quality-based clustering of gene expression profiles. *Bioinformatics*, 18(5):735–746.

Dembele, D. and Kastner, P. (2003). Fuzzy C-means method for clustering microarray data. *Bioinformatics*, 19(8):973–980.

D'Haeseleer, P. (2005). How does gene expression clustering work? *Nat. Biotechnol.*, 23:1499–1501.

Dopazo, J. and Carazo, J.M. (1997). Phylogenetic reconstruction using an unsupervised growing neural network that adopts the topology of a phylogenetic tree. *J. Mol. Evol.*, 44:226–233.

Dudoit, S. and Fridlyand, J. (2002). A prediction-based resampling method for estimating the number of clusters in a dataset. *Genome Biol.*, 3(7):RESEARCH0036.

Dudoit, S. and Fridlyand, J. (2003). Bagging to improve the accuracy of a clustering procedure. *Bioinformatics*, 19(9):1090–1099.

Eisen, M.B., Spellman, P.T., Brown, P.O., and Botstein, D. (1998). Cluster analysis and display of genome-wide expression patterns. *Proc. Natl. Acad. Sci. USA*, 95(25):14863–14868.

Ernst, J. and Bar-Joseph, Z. (2006). STEM: A tool for the analysis of short time series gene expression data. *BMC Bioinformatics*, 7:191.

Ernst, J., Nau, G.J., and Bar-Joseph, Z. (2005). Clustering short time series gene expression data. *Bioinformatics*, 21(1):i159–i168.

Ge, H., Walhout, A.J., and Vidal, M. (2003). Integrating 'omic' information: A bridge between genomics and systems biology. *Trends Genet.*, 19:551–560.

Gentleman, R.C., Carey, V.J., and Bates, D.M., et al. (2004). Bioconductor: Open software development for computational biology and bioinformatics. *Genome Biol.*, 5:R80.

Getz, G., Levine, E., and Domany, E. (2000). Coupled two-way clustering analysis of gene microarray data. *Proc. Natl. Acad. Sci. USA*, 97(22):12079–12084.

Ghosh, D. and Chinnaiyan, A.M. (2002). Mixture modelling of gene expression data from microarray experiments. *Bioinformatics*, 18(2):275–286.

Gibbons, F.D. and Roth, F.P. (2002). Judging the quality of gene expression-based clustering methods using gene annotation. *Genome Res.*, 12:1574–1581.

Golub, T.R., Slonim, D.K., Tamayo, P., Huard, C., Gaasenbeek, M., Mesirov, J.P., Coller, H., Loh, M.L., Downing, J.R., Caligiuri, M.A., Bloomfield, C.D., and Lander, E.S. (1999). Molecular classification of cancer class discovery and class prediction by gene expression monitoring. *Science*, 286(5439):531–537.

Hallikas, O., Palin, K., Sinjushina, N., Rautiainen, R., Partanen, J., Ukkonen, E., and Taipale, J. (2006). Genome-wide prediction of mammalian enhancers based on analysis of transcription-factor binding affinity. *Cell*, 124:47–59.

Handl, J., Knowles, J., and Kell, D.B. (2005). Computational cluster validation in post-genomic data analysis. *Bioinformatics*, 21(15):3201–3212.

Hastie, T., Tibshirani, R., Eisen, M.B., Alizadeh, A., Levy, R., Staudt, L., Chan, W.C., Botstein, D., and Brown, P. (2000). 'gene shaving' as a method for identifying distinct sets of genes with similar expression patterns. *Genome Biol.*, 1:RESEARCH0003.

Herrero, J., Al-Shahrour, F., Diaz-Uriarte, R., Mateos, A., Vaquerizas, J.M., Santoyo, J., and Dopazo, J. (2003). GEPAS: A web-based resource for microarray gene expression data analysis. *Nucleic Acids Res.*, 31:3461–3467.

Herrero, J., Valencia, A., and Dopazo, J. (2001). A hierarchical unsupervised growing neural network for clustering gene expression patterns. *Bioinformatics*, 17(2):126–136.

Herrero, J., Vaquerizas, J.M., Al-Shahrour, F., Conde, L., Mateos, A., Diaz-Uriarte, J.S., and Dopazo, J. (2004). New challenges in gene expression data analysis and the extended GEPAS. *Nucleic Acids Res.*, 32:W485–491.

Heyer, L.J., Kruglyak, S., and Yooseph, S. (1999). Exploring expression data: identification and analysis of coexpressed genes. *Genome Res.*, 9(11):1106–1115.

Horimoto, K. and Toh, H. (2001). Statistical estimation of cluster boundaries in gene expression profile data. *Bioinformatics*, 17(12):1143–1151.

Huang, D. and Pan, W. (2006). Incorporating biological knowledge into distance-based clustering analysis of microarray gene expression data. *Bioinformatics, in press*.

Hunter, L., Taylor, R.C., Leach, S.M., and Simon, R. (2001). GEST: A gene expression search tool based on a novel Bayesian similarity metric. *Bioinformatics, Supplement*, 17(1):S115–S122.

Ihmels, J., Friedlander, G., Bergmann, S., Sarig, O., Ziv, Y., and Barkai, N. (2002). Revealing modular organization in the yeast transcriptional network. *Nat. Genet.*, 31:370–377.

Jia, Z. and Xu, S. (2005). Clustering expressed genes on the basis of their association with a quantitative phenotype. *Genet. Res.*, 86:193–207.

Kanehisa, M., Goto, S., Kawashima, S., Okuno, Y., and Hattori, M. (2004). The KEGG resource for deciphering the genome. *Nucleic Acids Res.*, 32:D277–D280.

Kapushesky, M., Kemmeren, P., and Culhane, A.C., et al. (2004). Expression Profiler: Next generation–an online platform for analysis of microarray data. *Nucleic Acids Res.*, 32:W465–W470.

Kerr, M.K. and Churchill, G.A. (2001). Bootstrapping cluster analysis: assessing the reliability of conclusions from microarray experiments. *Proc. Natl. Acad. Sci. USA*, 98(16):8961–8965.

Khatri, P. and Draghici, S. (2005). Ontological analysis of gene expression data: Current tools, limitations, and open problems. *Bioinformatics*, 21(18):3587–3595.

Kohonen, T. (1997). *Self-Organizing Maps*. Springer, Berlin.

Kotlyar, M., Fuhrman, S., Ableson, A., and Somogyi, R. (2002). Spearman correlation identifies statistically significant gene expression clusters in spinal cord development and injury. *Neurochem. Res.*, 27:1133–1140.

Lazzeroni, L. and Owen, A. (2002). Plaid models for gene expression data. *Statistica Sinica*, 12:61–86.

Lee, H.K., Hsu, A.K., Sajdak, J., Qin, J., and Pavlidis, P. (2004). Coexpression analysis of human genes across many microarray data sets. *Genome Res.*, 14:1085–1094.

Mahalanobis, P. (1936). On the generalized distance in statistics. *Proc. Natl. Inst. Sci. India*, 12:49–55.

McLachlan, G.J., Bean, R.W., and Peel, D. (2002). A mixture model-based approach to the clustering of microarray expression data. *Bioinformatics*, 18(3):413–422.

McQueen, J. (1967). Some methods for classification and analysis of multivariate observations. *Proc. 5th Berkeley Symp. Math. Stat. Prob.*, pages 281–297.

Montaner, D., Tarraga, J., and Huerta-Cepas, J. (2006). Next station in microarray data analysis: GEPAS. *Nucleic Acids Res., in press*.

Moreau, Y., Aerts, S., de Moor, B., de Strooper, B., and Dabrowski, M. (2003). Comparison and meta-analysis of microarray data: From the bench to the computer desk. *Trends Genet.*, 19:570–577.

Pan, W. (2006). Incorporating gene functions as priors in model-based clustering of microarray gene expression data. *Bioinformatics*, 22(7):795–801.

Perou, C.M., Jeffrey, S.S., and van de Rijn, M., et al. (1999). Distinctive gene expression patterns in human mammary epithelial cells and breast cancers. *Proc. Natl. Acad. Sci. USA*, 96(16):9212–9217.

Prelic, A., Bleuler, S., and Zimmermann, P., et al. (2006). A systematic comparison and evaluation of biclustering methods for gene expression data. *Bioinformatics*, 22(9):1122–1129.

Quackenbush, J. (2001). Computational analysis of microarray data. *Nat. Rev. Genet.*, 2:418–427.

Ramoni, M.F., Sebastiani, P., and Kohane, I.S. (2002). Cluster analysis of gene expression dynamics. *Proc. Natl. Acad. Sci. USA*, 99(14):9121–9126.

Rousseeuw, P. (1987). Silhouettes: A graphical aid to the interpretation and validation of cluster analysis. *J. Comput. Appl. Math.*, 20:53–65.

Rual, J. F., Venkatesan, K., and Hao, T. (2005). Towards a proteome-scale map of the human protein-protein interaction network. *Nature*, 437:1173–1178.

Rudra, D., Zhao, Y., and Warner, J.R. (2005). Central role of Ifh1p-Fhl1p interaction in the synthesis of yeast ribosomal proteins. *EMBO J.*, 24:533–542.

Saeed, A.I., Sharov, V., and White, J. (2003). TM4: A free, open-source system for microarray data management and analysis. *Biotechniques*, 34:374–378.

Schliep, A., Schonhuth, A., and Steinhoff, C. (2003). Using hidden markov models to analyze gene expression time course data. *Bioinformatics, Suppl.*, 19(1):i255–i263.

Sheng, Q., Moreau, Y., and de Moor, B. (2003). Biclustering microarray data by Gibbs sampling. *Bioinformatics, Suppl.*, 19(2):II196–II205.

Simon, I., Siegfried, Z., Ernst, J., and Bar-Joseph, Z. (2005). Combined static and dynamic analysis for determining the quality of time-series expression profiles. *Nat. Biotechnol.*, 23:1503–1508.

Simon, R. (2005). Roadmap for developing and validation therapeutically relevant genomic classifiers. *J. Clin. Onc.*, 23(29):7332–7341.

Simon, R., Radmacher, M.D., Dobbin, K., and McShane, L.M. (2003). Pitfalls in the use of DNA microarray data for diagnostic and prognostic classification. *J. Natl. Cancer Inst.*, 95:14–18.

Slonim, D.K. (2002). From patterns to pathways: Gene expression data analysis comes of age. *The Chipping Forecast II, Nat. Gen.*, 32:502–508.

Sneath, P. and Sokal, R. (1973). *Numerical Taxonomy.* W.H. Freeman, San Francisco.

Spellman, P.T., Sherlock, G., Zhang, M.Q, Iyer, V.R., Anders, K., Eisen, M.B., Brown, P.O., Botstein, D., and Futcher, B. (1998). Comprehensive identification of cell cycle-regulated genes of the yeast *saccharomyces cerevisiae* by microarray hybridization. *Mol. Biol. Cell*, 9:3273–3297.

Stelzl, U., Worm, U., and Lalowski, M. (2005). A human protein-protein interaction network: A resource for annotating the proteome. *Cell*, 122:957–968.

Storey, J.D. and Tibshirani, R. (2003). Statistical significance for genomewide studies. *Proc. Natl. Acad. Sci. USA*, 100(16):9440–9445.

Stuart, J.M., Segal, E., Koller, D., and Kim, S.K. (2003). A gene-coexpression network for global discovery of conserved genetic modules. *Science*, 302:249–255.

Subramanian, A., Tamayo, P., and Mootha, V.K. (2005). Gene set enrichment analysis: A knowledge-based approach for interpreting genome-wide expression profiles. *Proc. Natl. Acad. Sci. USA*, 102(43):15545–15550.

Tanay, A., Sharan, R., and Shamir, R. (2002). Discovering statistically significant biclusters in gene expression data. *Bioinformatics, Suppl.*, 18(1):S136–S144.

Tavazoie, S., Hughes, J.D., Campbell, M.J., Cho, R.J., and Church, G.M. (1999). Systematic determination of genetic network architecture. *Nat. Genet.*, 22:281–285.

Toronen, P. (2004). Selection of informative clusters from hierarchical cluster tree with gene classes. *BMC Bioinformatics*, 5:32.

van't Veer, L.J., Dai, H., and van de Vijver, M.J., et al. (2002). Gene expression profiling predicts clinical outcome of breast cancer. *Nature*, 415:530–536.

Vaquerizas, J.M., Conde, L., and Yankilevich, P., et al. (2005). GEPAS, an experiment-oriented pipeline for the analysis of microarray gene expression data. *Nucleic Acids Res.*, 33:W616–W620.

Vaquerizas, J.M., Dopazo, J., and Diaz-Uriarte, R. (2004). DNMAD: Web-based diagnosis and normalization for microarray data. *Bioinformatics*, 20(18):3656–3658.

Vogl, C., Sanchez-Cabo, F., Stocker, G., Hubbard, S., Wolkenhauer, O., and Trajanoski, Z. (2005). A fully Bayesian model to cluster gene-expression profiles. *Bioinformatics, Suppl.*, 21(2):ii130–ii136.

Wrobel, G., Chalmel, F., and Primig, M. (2005). goCluster integrates statistical analysis and functional interpretation of microarray expression data. *Bioinformatics*, 21(17):3575–3577.

Wu, C.J. and Kasif, S. (2005). GEMS: A web server for biclustering analysis of expression data. *Nucleic Acids Res.*, 33:W596–W599.

Yeung, K.Y., Fraley, C., Murua, A., Raftery, A.E., and Ruzzo, W.L. (2001a). Model-based clustering and data transformations for gene expression data. *Bioinformatics*, 17(10):977–987.

Yeung, K.Y., Haynor, D.R., and Ruzzo, W.L. (2001b). Validating clustering for gene expression data. 17(4):309–318.

Zhu, G., Spellman, P.T., Volpe, T., Brown, P.O., Botstein, D., Davis, T.N, and Futcher, B. (2000). Two yeast forkhead genes regulate the cell cycle and pseudohyphal growth. *Nature*, 406:90–94.

Feature Selection and Dimensionality Reduction in Genomics and Proteomics

Milos Hauskrecht[1,2,3], Richard Pelikan[2], Michal Valko[1], and James Lyons-Weiler[3]

[1] Department of Computer Science, University of Pittsburgh, Pittsburgh, PA 15260, USA.
[2] Intelligent Systems Program, University of Pittsburgh, Pittsburgh, PA 15260, USA.
[3] Department of Biomedical Informatics, University of Pittsburgh, Pittsburgh, PA 15260, USA.
milos@cs.pitt.edu, pelikan@cs.pitt.edu, michal@cs.pitt.edu, lyonsweilerj@upmc.edu

7.1 Introduction

As technology improves, the amount of information we collect about the world increases. Sensor networks collect traffic or weather information in real-time, documents and news articles are distributed and searched on-line, information in medical records is collected and stored in electronic form. All of this information can be mined so that the relations among components of the underlying systems are better understood and their models can be built. Microarray and mass spectrometry (MS) technologies are producing large quantities of genomic and proteomic data relevant for our understanding of the behavior and function of an organism, or characteristics of disease and its dynamics. Thousands of genes are measured in a typical microarray assay; tens of thousands of measurements comprise a mass spectrometry proteomic profile. The high-dimensional nature of the data demands the development of special data analysis procedures that are able to adequately handle such data. The central question of this process becomes the identification of those *features* (measurements, attributes) that are most relevant for characterizing the system and its behavior. We study this problem in the context of classification tasks where our goal is to find features that discriminate well among classes of samples, such as samples from people with and without a certain disease.

Feature selection is a process that aims to identify a small subset of features from a large number of features collected in the data set. Two closely-related objectives may drive the feature selection process: (1) Building a reliable classification model which discriminates disease from control samples with high accuracy. The model is then applied to early detection and diagnosis of the disease. (2) Biomarker discovery task where a small set of features (genes in DNA microarrays, or peaks in proteomic spectra) that discriminate well between disease and control groups is identified so that the responsible features can be subjected to further laboratory exploration.

In principle, building a good classification model does not require feature selection. However, when the sample size is small in comparison to the number of features, feature selection may be necessary before a classification model can be reliably learned. With a small sample size, the estimates of parameters of the model may become unreliable and may cause *overfitting*, a phenomenon in which each datum is fit so rigidly that the model lacks flexibility for future data. To avoid overfitting, feature selection is applied to balance the number of features in proportion to the sample size. On the other hand, identification of a small panel of features for biomarker discovery purposes requires a classification model so that the discriminative behavior of the panel can be assessed.

The dimensionality of typical genomic and proteomic data sets one has to analyze surpasses the number of samples collected in typical studies by a large margin. For example, a typical microarray study can consist of up to a hundred samples with thousands of gene-expression measurements. Mass spectrometry (MS) proteomic profiling is less expensive and as a result one can often see data sets with two to three hundred profiles. MS profiles consist of thousands of measurements. Typically, "peaks" are selected among those measurements, and number in the hundreds. In either case, feature selection becomes important for both the biomarker discovery and interpretive analysis tasks; one has to seek a robust combination of feature selection methods and classification models to assure their reliability and success. Finally, feature selection may be a one-shot process, but typically, it is a search problem where more than one feature subset is evaluated and compared. Since the number of possible feature subsets is exponential in the number of constituent features, efficient feature selection methods are typically sought.

Feature selection methods are typically divided into three main groups: *Filter, wrapper* and *embedded methods*. Filter methods rank each feature according to some univariate metric, and only the highest ranking features are used; the remaining features are eliminated. Wrapper algorithms (Kohavi and John, 1998) search for the best subset of features. To assess the quality of a feature set, these methods rely on and interact with a classification algorithm and its ability to discriminate among the classes. The wrapper algorithm treats a classification algorithm as a black box, so any classification method can be combined with the wrapper. Standard optimization techniques (hill climbing, simulated annealing or genetic algorithms) can be used. Embedded methods search among different feature subsets, but unlike wrappers, the process is tied closely to a certain classification model and takes advantage of its characteristics and structure. In addition to feature selection approaches, in which a subset of original features is searched, the dimensionality problem can be often resolved via *feature construction*. The process of feature construction builds a new set of features by combining multiple existing features with the expectation that their combination improves our chance to discriminate among the classes as compared to the original feature space.

In this chapter, we first introduce the main ideas of four different methods for feature selection and dimensionality reduction and describe some of their representatives in greater depth. Later, we apply the methods to the analysis of one MS proteomic cancer data set. We analyze each method with respect to the quality of features selected and stress differences among the methods. Since our measuring criterion for feature effectiveness is how well it allows us to classify our samples, we compare the methods and their classification accuracy by combining them with a fixed classification method — a linear support vector machine (Vapnik, 1995). In closing, we analyze the results and give recommendations on the methods.

7.2 Basic Concepts

7.2.1 Filter Methods

Filter methods perform feature selection in two steps. In the first step, the filter method assesses each feature individually for its potential in discriminating among classes in the data. In the second step, features falling beyond some thresholding criterion are eliminated, and the smaller set of remaining features is used. This score-and-filter approach has been used in many recent publications, due to its relative simplicity. Scoring methods generally focus on measuring the differences between distributions of features. The resulting score is intended to reflect the quality of each feature in terms of its discriminative power. Many scoring criteria exist. For example, in the Fisher score (Pavlidis et al., 2001),

$$V(i) = \frac{(\mu_{(+)}(i) - \mu_{(-)}(i))^2}{\sigma^2_{(+)}(i) + \sigma^2_{(-)}(i)} \tag{7.1}$$

the quality of each feature is expressed in terms of the difference among the empirical means of two distributions, normalized by the sum of their variances. Table 7.1 displays examples of scoring criteria used in bioinformatics literature. Note that some of the scores can be applied directly to continuous quantities, while others require discretization. Scores can be limited to two classes, like the Fisher score, while others, such as the mutual information score, can be used in the presence of three or more classes. For the remainder of this chapter, we will assume our scoring metrics deal with binary decisions, where the data either belong to a positive (+) or negative (−) group.

7.2.1.1 Criteria Based on Hypothesis Testing

Some of the scoring criteria are related to statistical hypothesis testing and significance of their results. For example, the t-statistic is related to the null

Table 7.1. Examples of univariate scoring criteria for filter methods. See section Mathematical Details for definitions of these scores.

Criterion	References
Fisher score	(Golub et al., 1999; Furey et al., 2000; Pavlidis et al., 2001)
SAM scoring criterion	(Tusher et al., 2001; Storey and Tibshirani, 2003)
t-test	(Baldi and Long, 2001; Gosser, 1908)
Mutual information	(Tzannes and Noonan, 1973)
χ^2 (Chi square)	(Chernoff and Lehmann, 1954; Liu and Setiono, 1995)
AUC	(Hanley and McNeil, 1982)
$J5$ score	(Patel and Lyons-Weiler, 2004)

hypothesis H_0 under which the two class-conditional distributions $p(x|y = (+))$ and $p(x|y = (-))$ have the identical mean, that is $\mu_{(+)} = \mu_{(-)}$. The degree of violation of H_0 is captured by the p-value of the t-statistic with respect to the Student distribution. As a result, features can be ranked using the inverse of their p-value. Similarly, one can rank the features according to the inverse of the p-value of the Wilcoxon rank-sum test (Wilcoxon, 1945), a nonparametric method, testing the null hypothesis that the class-conditional densities of individual features are equal.

7.2.1.2 Permutation Tests

Any differential scoring metric (statistic) can be incorporated into and evaluated within the hypothesis testing framework via permutation tests. Permutation (or randomization) tests define a class of non-parametric techniques developed in the statistics literature (Kendall, 1945; Good, 1994), that are used to estimate the probability distribution of a statistic under the null (random) hypothesis from the available data. The estimate of the probability distribution of a scoring metric (Fisher score, J-measure, t-score, etc.) under the null condition allows us to estimate the p-value of the score observed in the data, similarly to the t-test or Wilcoxon rank-sum test. From the viewpoint of feature selection, the null hypothesis assumes that the conditional probability distributions for the two classes ($y = (+)$ or $(-)$) are identical under a feature x, that is, $p(x|y = (+)) = p(x|y = (-))$; or equivalently, that the data and the labels are independent, $p(x, y) = p(x)p(y)$. The distribution of data under the null hypothesis is generated through random permutations (of labels) in the data. The *permutation test algorithm* is shown below. The main cycle of the algorithm either scans through all possible permutations of labels, or, if this set is too large, a large number B of permutations is generated randomly. With sufficient cycles, the distribution of the test statistic under the null hypothesis can be estimated reliably.

---

permutation_test
```
{
    Compute the test statistic T for the original data;
    For b = 1 to B do
        {Permute randomly the group labels in the data;
         Compute the test statistic Tb for the modified data;
        }
    Calculate the p-value of T with respect to the distribution defined by
        permutations b as: p = N_{T_b ≥ T}/B;  where N_{T_b ≥ T} is the number of
        permutations for which the test statistic Tb is better than T;
    Return p;
}
```

7.2.1.3 Choosing Features Based on the Score

Differential scores or their associated p-value scores allow us to rank all feature candidates. However, it is still not clear how many features should be filtered out. The task is easy if we always seek a fixed set of k features. In such a case, the top k features are selected with respect to the ordering imposed by ranking features by their score. However, the quality of these features may vary widely, so selecting the features based solely on the order may cause some poor features to be included in the set. An alternative method is to choose features by introducing a threshold on the value of the score. Unfortunately, not every scoring criterion has an interpretable meaning, so it is unclear how to select an appropriate threshold. The statistic typically used for this purpose is the p-value associated with the hypothesis test. For example, if the p-value threshold is 0.05 then there is a 5% chance the feature is not differentially expressed at the threshold value. Such a setting allows us to control the chance of *false positive* selections. These are features which appear discriminative by chance.

7.2.1.4 Feature Set Selection and Controlling False Positives

The high-dimensional nature of biological data sources necessitates that many features (genes or MS-profile peaks) be tested and evaluated simultaneously. Unfortunately, this increases the chance that false positives are selected. To illustrate this, assume we measure the expression of 10 000 independent genes and none of them are differentially expressed. Despite the fact that there is no differential expression, we might expect 100 features to have their p-value smaller than 0.01. An individual feature with p-value 0.01 may appear good in isolation, but may become a suspect if it is selected from thousands of tested features. In such a case, the p-value of the combined set of the top 100 features selected out of 10 000 is quite different. Thus, adjustment of the p-value when performing multiple tests in parallel is necessary.

The *Bonferroni correction* adjusts the p-value for each individual test by dividing the target p-value for all findings by the number of findings. This assures that the probability of falsely rejecting any null hypotheses is less than or equal to the target p. The limitation of the Bonferroni correction is that it operates under the assumption of independence and as a result it is too conservative if features are correlated. Two alternatives to the Bonferroni correction are offered by: (1) the *family-wise error rate method* (FWER, (Westfall and Young, 1993)) and (2) methods for controlling the *false discovery rate* (FDR, (Benjamini and Hochberg, 1995; Tusher et al., 2001). FWER takes into account the dependence structure among features, which often translates to higher power. Benjamini and Hochberg (1995) suggest to control FDR instead of the p-value. The FDR is defined as the mean of the number of false rejections divided by the total number of rejections. The *significance analysis of microarrays* (SAM) method (Storey and Tibshirani, 2003) is used as an estimate of the FDR. Depending on the chosen threshold value for the test statistic T, it estimates the expected proportion of false positives on the feature list using a permutation scheme.

7.2.1.5 Correlation Filtering

To keep the feature set small, the objective is to diversify the features as much as possible. The selected features should be discriminative as well as independent from each other as much as possible. The rationale is that two or more independent features will be able to discriminate the two classes better than any of them individually. Each feature may differentiate different sets of data well, and independence between the features tends to reduce the overlap of the sets. Similarly, highly dependent features tend to favor the same data and thus are less likely to help when both are included in the panel. The extreme case is when the two features are exact duplicates, in which case one feature can be eliminated.

Correlation filters (Ross et al., 2000; Hauskrecht et al., 2005) try to remove highly correlated features since these are less likely to add new discriminative information (Guyon and Elisseeff, 2003). Various elimination schemes are used within these filters to reduce the chance of selected features being highly correlated. Typically, correlation filters are used in combination with other differential scoring methods. For example, features can be selected incrementally according to their p-value; the feature to be added next is checked for correlation with previously selected features. If the new feature exceeds some correlation threshold, it is eliminated (Hauskrecht et al., 2005).

7.2.2 Wrapper Methods

Wrapper methods (Kohavi and John, 1998) search for the best feature subset in combination with a fixed classification method. The goodness of a feature subset is determined using internal-validation methods, such as, k-fold or leave-one-out cross-validation (Krus and Fuller, 1982). Since the number of all combinations is exponential in the number of features, the efficiency of the search methods is often critical for its practical acceptance. Different heuristic optimization frameworks have been applied to search for the best subset. These include: *Forward selection, backward elimination* (Blum and Langley, 1997), *hill climbing, beam search* (Russel and Norvig, 1995), and randomized algorithms such as *genetic algorithms* (Koza, 1995) or *simulated annealing* (Kirkpatrick et al., 1983). In general, these methods explore the search space (subsets of all features) starting with no features, all features, or a random selection of features. For example, the forward selection approach builds a feature set by starting from an empty feature set and incrementally adding the feature that improves the current feature set the most. The procedure stops when no improvement in the feature set quality is possible.

7.2.3 Embedded Methods

Embedded methods incorporate variable selection as part of the model building process. A classic example of an embedded method is CART (Classification and Regression Trees, (Breiman et al., 1984)).

CART searches the range of each individual feature to find the split that optimally divides the observed data into a more homogeneous groups (with respect to the outcome variable). Beginning with the subsets of the variable that produces the most homogeneous split, each variable is again searched across its range to find the next optimal split. This process is continued within each new subset until all data are perfectly fit by the resulting tree, or the terminal nodes have a small sample size. The group constituting the majority of data points in each node determines the classification accuracy of the derived terminal nodes. Misclassification error from internal cross-validation can be used to backprune the decision tree and optimize its projected generalization performance on additional independent test examples.

7.2.3.1 Regularization/Shrinkage Methods

Regularization or shrinkage methods (Hastie et al., 2001; Xing et al., 2001) offer an alternative way to learn classifications for data sets with large number of features but small sample size. These methods trim the space of features directly during classification. In other words, regularization "effectively" shuts down (or zeros the influence of) unnecessary features.

Regularization can be incorporated either into the error criterion or directly into the model. Let **w** be a set of parameters defining a classification model (e.g., the weights of a logistic regression model), and let $Error(\mathbf{w}, \mathbf{D})$ be an error function reflecting the fit of the model to data (e.g., least-squares as likelihood-based error). A regularized error function is then defined as:

$$Error_{Reg}(\mathbf{w}, \mathbf{D}) = Error(\mathbf{w}, \mathbf{D}) + \lambda \|\mathbf{w}\|, \qquad (7.2)$$

where $\lambda > 0$ is a regularization constant, and $\| \cdot \|$ is either the L_1 or L_2 norm. Intuitively, the regularization term penalizes the model for nonzero weights so the optimization of the new error function drives all unnecessary parameters to 0. Automatic relevance determination (ARD) (MacKay, 1992; Neal, 1998) achieves regularization effects in a slightly different way. The relevance of an individual feature is represented explicitly via model parameters and the values of these parameters are learned through Bayesian methods. In both cases, the output of the learning is a feature-restricted classification model, so features are selected in parallel with model learning.

7.2.3.2 Support Vector Machines

Regularization effects are at work also in one of the most popular classification frameworks these days: The support vector machine (SVM) (Burges, 1998; Schölkopf and Smola, 2002). The SVM defines a linear decision boundary (hyperplane) that separates case and control examples. The boundary maximizes the distance (also called *margin*) in between the two sample groups. The effects of margin optimization are twofold: Only a small set of data points (support vectors) are critical for the separation; the dimensions unnecessary for separation are penalized. Both of these processes help to fight the problem of model overfit. As a result, the SVM offers a robust classification framework that works very well for situations with a moderately large number of features and relatively small sample sizes.

7.2.4 Feature Construction

Better discriminatory performance can be often achieved using features constructed from the original input features. Building a new feature is an opportunity to incorporate domain specific knowledge into the process and hence to improve the quality of features. Nevertheless, a number of generic feature construction methods exist: Clustering; linear (affine) projections of the original feature space; as well as more sophisticated space transformations such as wavelet or kernel transforms. In the following, we briefly review three basic feature construction approaches: Clustering, PCA and linear discriminative projections.

7.2.4.1 Clustering

Clustering groups data components (data points or features) according to their similarity. Every data component is assigned to one of the groups (clusters); components falling into the same cluster are assigned the same value in the new (reduced) representation. Clustering is typically used to identify distinguished sample groups in data (Ben-Dor et al., 2000; Slonim et al., 2000). In contrast to supervised learning techniques that rely heavily on class label information, clustering is unsupervised and the information about the target groups (classes) is not used. From the dimensionality reduction perspective, a data point is assigned a cluster label which is then used as its representation.

Clustering methods rely on the similarity matrix – a matrix of distances between data components. The similarity matrix can be built using one of the standard distance metrics such as Euclidean, Mahalanobis, Minkowski, etc., but more complex distances based on, for example, functional similarity of genes (Speer et al., 2005), are possible. Table 7.2 gives a list of some standard distance metrics one may use in clustering.

Table 7.2. Examples of distance metrics for clustering.

Metric	Formula
Euclidean distance	$d(r,s) = \sqrt{(\mathbf{x}_r - \mathbf{x}_s)(\mathbf{x}_r - \mathbf{x}_s)'}$
Standardized Euclidean distance	$d(r,s) = \sqrt{(\mathbf{x}_r - \mathbf{x}_s)D^{-1}(\mathbf{x}_r - \mathbf{x}_s)'}$
Mahalanobis distance	$d(r,s) = \sqrt{(\mathbf{x}_r - \mathbf{x}_s)\Sigma^{-1}(\mathbf{x}_r - \mathbf{x}_s)'}$
City Block (or Manhattan) metric	$d(r,s) = \sum_{j=1}^{n} |\mathbf{x}_{rj} - \mathbf{x}_{sj}|$
Minkowski metric	$d(r,s) = \sqrt[p]{\left(\sum_{j=1}^{n} |\mathbf{x}_{rj} - \mathbf{x}_{sj}|^p\right)}$
Cosine distance	$d(r,s) = \left(1 - \frac{\mathbf{x}_r \mathbf{x}_s'}{\sqrt{\mathbf{x}_r' \mathbf{x}_r}\sqrt{\mathbf{x}_s' \mathbf{x}_s}}\right)$
Correlation distance	$d(r,s) = 1 - \frac{(\mathbf{x}_r - \bar{\mathbf{x}}_r)(\mathbf{x}_s - \bar{\mathbf{x}}_s)'}{\sqrt{(\mathbf{x}_r - \bar{\mathbf{x}}_r)(\mathbf{x}_r - \bar{\mathbf{x}}_r)'}\sqrt{(\mathbf{x}_s - \bar{\mathbf{x}}_s)(\mathbf{x}_s - \bar{\mathbf{x}}_s)'}}$
Hamming distance	$d(r,s) = \frac{\#(x_{rj} \neq x_{sj})}{n}$
Jaccard distance	$d(r,s) = \frac{\#[(x_{rj} \neq x_{sj}) \wedge ((x_{rj} \neq 0) \vee (x_{sj} \neq 0))]}{\#[(x_{rj} \neq 0) \vee (x_{sj} \neq 0)]}$

$\mathbf{x}$ and $\mathbf{x}'$ denote a column vector and its transpose, respectively.
$\mathbf{x}_r$ and $\mathbf{x}_s$ indicate the rth and s^{th} samples in the data set, respectively.
x_{rj} indicates the jth feature of the rth sample in the data set.
$\bar{\mathbf{x}}_r$ indicates the mean of all features in the rth sample in the data set.
D is the diagonal matrix with diagonal elements given by v_i^2, which denotes the variance of ith variable.
Σ is the sample covariance matrix.
The symbol $\#$ denotes counts; the number of instances satisfying the associated property.

7.2.4.2 Clustering Algorithms

The goal of clustering is to optimize intra- and inter-cluster distances among the components. Two basic clustering algorithms are: *k-means clustering* (McQueen, 1967; Ball and Hall, 1967), and *hierarchical agglomerative clustering* (Cormack, 1971; Eisen et al., 1998).

Briefly, the *k*-means algorithm clusters data into groups by iteratively optimizing positions of cluster centers (means) so that the sum of within-cluster distances (the distances between data points and their cluster centers) is minimized. Initial positions for cluster centers are generated randomly or by using heuristics. The algorithm is not guaranteed to converge to the optimal solution. On the other hand, hierarchical agglomerative methods work by combining pairs of data entities (features) or clusters into a hierarchical structure (called a dendrogram). The algorithm starts from unit clusters and merges them greedily (i.e., choosing the merge which most improves the fit of the clusters to the data) into larger clusters using an a priori selected similarity measure.

7.2.4.3 Probabilistic (Soft) Clustering

The *k*-means and agglomerative clustering methods assign every data point into a single cluster. However, sometimes it may be hard to decide what cluster the point belongs to. In *probabilistic (soft) clustering* methods, a data point belongs to all clusters, but the strength (weight) of its association with clusters differs by how well it fits cluster descriptions. Typically, the weight has probabilistic meaning and defines a probability with which a data point belongs to a cluster.

To calculate the probability, an underlying probabilistic model must be first fit to the data. Briefly, data are assumed to be generated from k different classes that correspond to clusters. Each class has its own distribution for generating data points. The parameters of these distributions as well as class (cluster) priors are fit (learned) using Expectation-Maximization techniques (Dempster et al., 1977). Once the model parameters are known, the probabilistic weights relating a data point and clusters are posterior probabilities of the point belonging to classes. A classic example of a probabilistic model often used in clustering is the Mixture of Gaussians model (McLachlan et al., 1997), where k clusters are modeled using k Gaussian distributions.

7.2.4.4 Clustering Features

Clustering methods can be applied to group either data points or features in the data. When clustering features, the dimensionality reduction is achieved by selecting a representative feature (typically the feature that is closest to the cluster center (Guyon and Elisseeff, 2003)), or by aggregating all features within the cluster via averaging to build a new (mean) feature. If we assume k different feature clusters, the original feature space is reduced to a new k-dimensional space. An example method of feature clustering is to cluster features based on intra-correlation, and use the cluster center as a representative. Closely correlated features are not likely to help when separated, so grouping them away from more unrelated features will help diversify the resulting features.

7.2.4.5 Principal Component Analysis

Principal component analysis (PCA) (Jolliffe, 1986) is a widely used method for reducing the dimensionality of data. PCA finds projections of high-dimensional data into a lower dimensional subspace such that the variance retained in the projected data is maximized. Equivalently, PCA gives uncorrelated linear projections of data while minimizing their least square reconstruction error. Additionally, PCA works fully unsupervised; class labels are ignored. PCA can be extended to nonlinear projections using kernel methods (Bach and Jordan, 2001). Dimensionality reduction methods similar to PCA that let us project high dimensional features into a lower dimensional space include multidimensional scaling (MDS) (Cox and Cox, 1994) used often for data visualization purposes or independent component analysis (ICA) (Jutten and Herault, 1991).

7.2.4.6 Discriminative Projections

Principal component analysis identifies affine (linear) projections of data that maximize the variance observed in data. The method operates in a fully unsupervised manner; no knowledge of class labels is used to find the principal projections. The question is whether there is a way to identify linear projections of features such that they optimize the discriminability among the two classes. Techniques which try to achieve this goal include *Fisher's linear discriminant* (FLD) (Duda et al., 2000), *linear discriminant analysis* (Hastie et al., 2001) and more complex methods like *partial least squares* (PLS) (Denham, 1994; Dijkstra, 1983).

   Take, for example, the linear discriminant analysis model. The model assumes that cases and controls are generated from two Gaussian distributions with means $\mu_{(-)}$, $\mu_{(+)}$ and the same covariance matrix Σ. The parameters of the two distributions are estimated from data using the maximum likelihood methods. The decision boundary that is defined by data points that give the same probability for both distributions is a line. The linear projection is defined as:

$$w = \Sigma^{-1}(\mu_{(+)} - \mu_{(-)}), \tag{7.3}$$

where $\mu_{(-)}, \mu_{(+)}$ are the means of the two groups and Σ is the covariance for both groups, where $p(x|y) \sim N(\mu, \Sigma)$.

7.3 Advantages and Disadvantages

Each of the aforementioned methods comes with advantages and disadvantages. The following text briefly summarizes them.

Filter methods:

- Advantages: Univariate scores are very easy to calculate and thus, filter methods have a short running time. If our goal is a prediction, they often perform well in combinations with more robust classification methods such as the SVM.
- Disadvantages: Many differential scoring methods exist, it is unclear which one is best for the data set at hand. The features are analyzed independent of each other. This is a problem if our goal is to identify a small panel of discriminative features (biomarkers). Multivariate relations/dependencies must be incorporated through additional criteria, e.g., correlation filters.

Wrapper methods:

- Advantages: More comprehensive search of the feature set space. The feature set with the best discriminative potential on a fixed classification method is selected.
- Disadvantages: Running time is much longer than filter methods; many feature sets need to be analyzed and assessed. In addition, scoring of feature sets is based on internal cross-validation methods, which lengthens their running time. The reliability of the estimate of the internal cross-validation error needs to be considered. Low reliability of the internal validation error in combination with a large number of subsets examined can be lethal especially in various greedy search schemes.

Embedded methods:

- Advantages: Features and their selection are tuned to a specific model. Learning methods which incorporate aspects of regularization, like the SVM or regularized logistic regression, can learn very good predictive models even in the presence of high-dimensional data. We recommend trying SVM as a first step if the goal is only to build a predictive model.
- Disadvantages: Identification of a small set of features may be problematic. Backward feature elimination routines (Guyon and Elisseeff, 2003) can be used to reduce the feature panel to a more reasonable size.

Feature construction methods:

- Advantages: May incorporate the domain knowledge which may translate to improved feature sets.
- Disadvantages: If features are constructed using one of the out-of-box methods (e.g., PCA) the new features may be hard to interpret biologically. In addition, many feature construction techniques (e.g., clustering, PCA, ICA) work in an unsupervised mode, so high-quality features for discriminatory purposes are not guaranteed.

7.4 Case Study: Pancreatic Cancer

To illustrate some of the advantages and disadvantages of feature selection methods, we use a data set of MS proteomic profiles for pancreatic cancer collected at the University of Pittsburgh Cancer Institute (UPCI). Since full feature selection comparison is very hard to do without a full predictive model that combines both the feature selection and the classification stages we test feature selection methods in combination with one classification method – the linear support vector machine (SVM) (Vapnik, 1995). All classification results presented in the following text were obtained by using the repeated random subsampling strategy with 40 different train/test data splits using 70/30 train/test split ratio. The optimization criterion for the SVM method was a zero-one loss function, which focuses on improving classification error instead of sensitivity or specificity. The statistics reported are: Average test classification error (ACE), sensitivity (SN) and specificity (SP) and their standard deviations.

7.4.1 Data and Pre-Processing

The data set consists of 116 MS profiles, with 57 cancer cases (+ group) and 59 controls, matched according to their smoking history, age, and gender (– group). The data were generated using Ciphergen Biosystems Inc. SELDI-TOF (surface-enhanced laser desorption/ionization time-of-flight) mass spectrometry. Compounds such as proteins, peptides and nucleic acids for masses of up to 200 000 Daltons are recorded using this technology. Before applying feature selection techniques the data set was pre-processed using the *Proteomic Data Analysis Package* (PDAP) (Hauskrecht et al., 2005). The following pre-processing steps were applied: (1) Cuberoot variance stabilization, (2) local min-window baseline correction, (3) Gaussian kernel smoothing, (4) range-restricted intensity normalization, and (5) peak-based profile alignment. The quality of all profiles were tested beforehand on raw MS profile readings using total ion current (TIC). None of the profiles differed by more than two standard deviations from the mean TIC, which is our current quality-assurance/quality–control threshold for sample exclusion. After basic pre-processing, peaks in the range of 1 500 – 1 650 Daltons were identified and their corresponding intensities were extracted.[4] This gave us a data set of 116 samples with 602 peak features.

---

[4] The region below 1 500 Daltons is unsuitable for analysis because of known signal reproducibility problems. The region is often referred to as the junk region. On the other hand, signals for higher mass-to-charge-ratios are of lower intensity which makes them hard to separate from the noise. An a priori upper limit is typically set to restrict the search for signal.

7.4.2 Filter Methods

7.4.2.1 Basic Filter Methods

Many univariate scoring metrics that assess the individual quality of features were proposed in the literature. An important question is how the rankings and subsequent feature selection induced by these metrics vary. Table 7.3 shows the number of overlapping features for the top 20 features selected according to four frequently used scoring criteria: Correlation, Fisher, t-statistic and Wilcoxon's p-value measures.

Table 7.3. Overlap of top 20 features for four different metrics.

	Correlation	Fisher	t-statistics	Wilcoxon
Correlation	–	18	12	18
Fisher	18	–	11	16
t-statistics	12	11	–	11
Wilcoxon	18	16	11	–

The table shows that different scoring metrics may induce rather different feature orders and as a result, different feature panels. It is very hard to argue that any one of them is the best. The quality depends strongly on the classification technique used in the next step, but even there the story is often unclear, and the best method tends to vary among the data sets. Table 7.4 illustrates the results obtained using top 20 choices of four scoring methods from Table 7.3 after we combine them with the linear SVM model. Standard deviations of performance statistics are also given. We see that the best classification error was obtained using the features selected based on the t-statistic score. While our experience is that the t-statistic score performs well on many proteomic data sets, other scoring metrics may often outperform it.

Table 7.4. Results for classifiers based on different feature filtering methods and the linear SVM. Standard deviations are given in parentheses.

	Correlation	Fisher	t-statistics	Wilcoxon
ACE	0.2500 (0.1178)	0.2188 (0.1075)	0.1743 (0.0684)	0.2611 (0.1091)
SN	0.8022 (0.0945)	0.8102 (0.1210)	0.8259 (0.0997)	0.7956 (0.1200)
SP	0.7142 (0.1249)	0.7628 (0.1423)	0.8327 (0.0852)	0.6961 (0.1607)

7.4.2.2 Controlling False Positive Selections

A problem with high-dimensional data is that some features may appear as good discriminators simply by chance. The problem of false positive identifications of features is critical for the biomarker discovery task. Clearly, a more comprehensive analysis and validation of the feature in the lab may incur a significant monetary cost. While positive feature selections may influence also the generalizations of the predictive model and its classification accuracy, the classification methods are often more robust to handle them and the problem of false positive features is less pressing than for the biomarker discovery applications.

The false positive selection rate can be controlled via p-value on individual features, Bonferroni corrected p-value for the panel of features, or through false discovery rate. Table 7.5 shows the number of features out of 602 original features selected by each of these methods.

Table 7.5. P-value for t-statistics.

original number of features	$p < 0.05$	Bonferroni $p < 0.05$	FDR 0.2
602	13	0	5

Assuming that all features are independent and random, we expect to see about 30 false positive features under the simple p-value of 0.05 for each feature. Using this estimate and the fact that we see only 13 features for the p-value of 0.05 would lead us to the conclusion that all of these are likely obtained by chance. The caveat is that when features are dependent and correlated the expected numbers are very different. Indeed, features in this and other proteomic data sets exhibit a large amount of correlation among the features; so the result in the table is indicative of such a dependency. The Bonferroni correction typically leads to a very conservative bound that may be very hard to satisfy. For example, none of the features in our cancer data passed Bonferroni-corrected p-value of 0.05. FWER and FDR methods and their thresholds give better estimates of false positive selections and their rates for the real-world data and should be preferred over simple and Bonferroni-corrected p-value thresholding.

When selecting features, our objective is to strike the right balance between the number of features, the flexibility they may offer when building multivariate discriminators, and the risk of inclusion of false positive features. The FWER and FDR methods give better control over risks of false positives. However, choosing the optimal thresholds for these techniques is a matter of personal preference. For example, two different approaches can be taken. If the selected features are meant only for use with an automated classification routine, it may be more acceptable to risk selecting false positives, and thusly the threshold can be less stringent. On the other hand, if the selected features are to be investigated more thoroughly (e.g., to analyze them using wet lab techniques), it would be far less acceptable to suggest that false positives are informative features. In this case, the threshold should be set more aggressively.

7.4.2.3 Correlation Filters

Biological (genomic and proteomic) data sets often exhibit a relatively high number of correlations. The correlations can be introduced by the technology producing the data or they reflect true underlying dependencies among measured species. For example, a peak in a proteomic profile is formed by a collection of correlated measurements, triple or double charged ions cause the same signal to be replicated at different parts of the profiles, and finally some peaks are correlated because they share a common regulatory (or interaction) pathway.

Selecting two features that are near duplicates, even if they are highly discriminative, does not help the classification model and its accuracy. Correlation filtering alleviates the problem by removing features highly correlated with existing features in the panel. Table 7.6 illustrates the number of features one obtains by filtering out correlated features at different maximum allowed absolute correlation (MAC) thresholds from the original 602 features. We note that the amounts of correlates filtered out at higher thresholds are statistically significantly different (at $p = 0.01$) from what one would obtain for independent feature sets.

Table 7.6. Effect of correlation filtering.

Threshold	1	0.9	0.8	0.7	0.6	0.5	0.4	0.3	0.2	0.1	0
Number of Features	602	460	247	119	52	22	12	9	6	3	1

Figure 7.1 illustrates the effect of correlation filtering when it is combined with the univariate feature scoring based on the t-statistic. We see that test errors for smaller feature sets (size 5) are improved if feature panels are decorrelated. However, for larger feature panels the effect of feature decorrelation may vanish since some good features that add some discriminative value to the panel are filtered out. For example, for 20 features in Figure 7.1 the effect of correlation filtering has disappeared and the SVM classifier based on the unrestricted t-statistic score performs better than classifiers with correlation thresholds of 0.75 and 0.5. This illustrates one of the problems of the method, identification of an appropriate MAC threshold. We must note that the effect as seen in Figure 7.1 may be less pronounced on other classification methods or on other data sets, while in some cases correlation thresholds may lead to superior performance. These differing outcomes are the results of tradeoffs of feature quality and overfit processes.

The plain correlation threshold filtering method suffers from a couple of problems. First, an identification of an appropriate correlation threshold in advance is hard. Moreover, for different feature sizes there appears to be a different threshold that works best so switching of thresholds may be appropriate. One solution to this problem is the parallel correlation filtering method

(Hauskrecht et al., 2005) that works at multiple correlation threshold levels in parallel and uses internal cross-validation methods to decide on what feature (correlation level) to select next. The performance of the method is compared to the unrestricted t-statistic filter and two correlation filtering methods based on simple MAC thresholds in Figure 7.1.

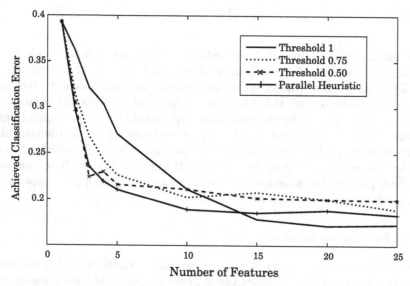

Fig. 7.1. Effect of correlation filtering on classification errors. Results of correlation filtering on the t-statistic score and SVM are shown.

7.4.3 Wrapper Methods

Wrapper methods search for the best subset of features by trying them in combination with a fixed classification method. However, there is a natural tradeoff between the quality of the feature set found, and the time taken to search for it. Table 7.7 displays performance statistics for two search methods: *Greedy forward selection* and *simulated annealing*.

The forward selection approach, also called the *greedy* approach, adds the feature which improves the set the most. The panel begins empty and is built incrementally, stopping when no improvement in the feature set is possible. Simulated annealing is a randomized algorithm and if it is left to search long enough all possible combinations may be reached and evaluated. Thus, simulated annealing may arrive at a better solution than the greedy method when given enough time. This quality/time tradeoff is captured in the table. The model based on the greedy forward selection method leads to average errors of 0.1750 while simulated annealing approaches 0.1660. To reach

Table 7.7. Wrapper methods with two search algorithms: Forward selection and simulated annealing. Standard deviations are given in parentheses.

	Greedy	Simulated Annealing
ACE	0.1750 (0.0668)	0.1660 (0.0603)
SN	0.8239 (0.1123)	0.8149 (0.1097)
SP	0.8261 (0.1100)	0.8614 (0.0784)
# steps	7037.4	10000

the result, 7 037.4 feature sets were evaluated on average by forward selection, while simulated annealing was run for 10 000 steps on every train/test split.

Evaluating a new feature set in any wrapper method is done by internal validation methods, such as k-fold cross-validation or leave-one-out validation. The overhead incurred by the evaluation step contributes to the running time of the algorithm. In general, using more internal splits improves the estimate of the error for each feature set. The price paid for it is an additional increase in the running time. Despite the downfalls, the results obtained from wrapper methods powered by various search heuristics are often quite good, especially when computational time is not an issue.

7.4.4 Embedded Methods

Table 7.8 shows the results of three classification methods with embedded feature selection: CART (Breiman et al., 1984), *regularized logistic regression* (RLR) (Hastie et al., 2001) and *support vector machines* (SVMs) (Burges, 1998). Each of these methods handles features differently, and consequently leads to different classification accuracies. We see that two of the methods, RLR and SVM, achieved results comparable or better than filter and wrapper methods. While this is not the rule, the linear SVM appears to be a very stable method across a large range of features so we always recommend to try it on the full feature set.

Table 7.8. Performance statistics for embedded methods. Standard deviations are given in parentheses.

	CART	Regularized LR	SVM
ACE	0.3681 (0.0897)	0.1382 (0.0584)	0.1382 (0.0623)
SN	0.6321 (0.1888)	0.8619 (0.1026)	0.8536 (0.0913)
SP	0.6361 (0.2088)	0.8624 (0.0942)	0.8769 (0.0881)

Embedded methods may not be optimal, if we want to use them for biomarker discovery, that is, if our objective is to find a small set of original features with a good discriminatory performance. The embedded methods may rely on too many features so a follow-up selection of a smaller subset is necessary. Wrapper methods based on the backward feature elimination (Guyon and Elisseeff, 2003) achieve this by gradually eliminating the features that affect the performance the least.

7.4.5 Feature Construction Methods

To illustrate feature construction methods we use three unsupervised methods: sample clustering, feature clustering and PCA projections, all aimed to reduce the dimensionality of data. The results of these methods in combination with the linear SVM are in table 7.9.

Table 7.9. Construction methods: Sample clustering using squared Euclidean distance, feature clustering using correlation coefficient, and PCA. Standard deviations are given in parentheses.

	Sample Clustering	Feature Clustering	PCA Projections
ACE	0.4525 (0.0810)	0.2104 (0.0652)	0.1681 (0.0594)
SN	0.4721 (0.1604)	0.7932 (0.1426)	0.8223 (0.0984)
SP	0.6444 (0.1633)	0.7968 (0.0920)	0.8492 (0.0842)

The first entry in the table (sample clustering with Euclidean distance) illustrates the major weakness of clustering methods: The clustering does not give reasoning as to why the data components group together, other that their distance is close, which obviously depends on the choice of the metric. Thus, one has to assure that the distance selected is not arbitrary and makes sense for the data and the prediction task. The result for clustering of features based on the correlation metric also supports this point. There are many feature correlates in the proteomic data set, so grouping the features based on their mutual correlation and replacing the features in each cluster with a feature corresponding to the cluster center tends to eliminate high correlates in the new (reduced) data. This is very similar in spirit to the correlation filtering method. The difference is that the correlation filtering is closely combined with and benefits from the univariate score filtering, while correlation clustering works fully unsupervised.

PCA constructs features using linear projections of complete data. Since PCA arranges projections along uncorrelated axes, it helps to relieve us from identifying feature correlates. As a result, we see an improvement in classification error over some other construction and filtering methods. Note that PCA can be a good "one shot" technique, avoiding necessities like the choice of the number of clusters, k, in k-means clustering, or scoring metric in filtering methods. The effort saved by not choosing parameters is in exchange for knowledge about a targetable panel of biomarkers, but PCA can still be convenient if the only interest is constructing a predictive model.

7.4.6 Summary of Analysis Results and Recommendations

There are multiple feature selection/dimensionality reduction methods one may apply to reduce the feature size of the data and make it "comparable" to its sample size. Unfortunately, there is no perfect recipe for what method to choose but here are some guidelines.

- Having prior information about how features can be related to the prediction task will always help feature selection and its subsequent application. So whenever possible try to use this information. For example, when the biological relevance of features can be ascertained, the potentially irrelevant or obvious features can also be eliminated.
- In the presence of no prior information, more generic information can be used for steering feature selection in the right direction. The effect of a feature on the target class and the presence of multivariate dependencies (e.g., correlations) among feature candidates appear to be the most important ones. The importance of a feature is captured by a univariate scoring metric. Dealing with highly correlated features, either by grouping them or eliminating redundancies, can help the selection process by narrowing the choice of features.
- Feature selection coupled with more robust classification methods, like SVM, can perform extremely well on all features. Backward feature elimination methods can be applied if we would like to identify a smaller panel of informative features.
- The feature selection method applied to data does not have to match a single method. A combination of feature selection methods may be beneficial and may work much better (Xing et al., 2001). For example, it may help to exclude some features outright with a basic filtering method by removing the lowest-scoring features and apply other methods (e.g., wrapper or PCA methods) only on the remaining features.

Since there are many feature selection methods, one may be tempted to try many of them in combination with a specific classifier and pick the one that gives the best test set result *post hoc*. Note that in such a case the error is biased and does not objectively report on the generalizability of the approach. Model selection methods based, for example, on an internal cross-validation loop should be applied whenever a choice out of many candidates is allowed.

In closing, it is important to note that the selection of the feature selection technique should first be driven by prior knowledge about the data, and then by the primary goal you wish to accomplish by analyzing the data: Obtain a small, easy to interpret, feature panel or build a good classification model. Feature selection techniques vary in their complexity and interpretability, and the issues discussed above must be taken into careful consideration.

7.5 Conclusions

In this chapter, we have presented four basic approaches to feature selection and dimensionality reduction. Filter, wrapper, and embedded methods work with the available features and choose those which appear important. In slight contrast, feature construction methods build new features which can be more powerful than previous ones. To discuss the entire gamut of feature selection methods would be exhaustive, as researchers must constantly meet their needs of analyzing high-dimensional data. The techniques covered here are among the most effective for analyzing genomic and proteomic data, in terms of building predictive models and developing biologically relevant information.

7.6 Mathematical Details

Table 7.10. Formulae for popular filter scores.

Filter Name	Formula
Fisher score	$score(i) = \frac{(\mu_+(i)-\mu_-(i))^2}{s_+^2(i)+s_-^2(i)}$
SAM score	$score(i) = \frac{|\mu_+(i)-\mu_-(i)|}{s_{SAM}(i)+s_{SAM,0}}$
t-test	$score(i) = \frac{|(\mu_+(i)-\mu_-(i))|}{\sqrt{\frac{s_+^2(i)}{n_+}+\frac{s_-^2(i)}{n_-}}}$
Mutual Inf.	$score(i) = \sum_{\{x_i\}} \sum_{y\in\{+,-\}} p(X_i=x_i, Y=y) \cdot \log \frac{p(X_i=x_i,Y=y)}{p(X_i=x_i)\cdot(Y=y)}$
χ^2 (Chi-square)	$score(i) = \sum_{\{x_i\}} \sum_{y\in\{+,-\}} \frac{(p(X_i=x_i,Y=y)-p(X_i=x_i)\cdot p(Y=y))^2}{p(X_i=x_i)\cdot p(Y=y)}$
AUC	$score(i) = $ Area under the ROC curve for feature i
J5 score	$score(i) = \frac{|\mu_+(i)-\mu_-(i)|}{\frac{1}{m}\sum_{j=1}^{m}|\mu_+(j)-\mu_-(j)|}$

where $\mu(i)$ and $s^2(i)$ denote the sample mean and sample variance of the ith feature, respectively. The signs $+$ and $-$ denote positive and negative examples, respectively.

The SAM technique is meant to be used in a permutation setting, however, the employed statistics can still be used for feature filtering. The terms $s_{SAM}(i)$ and $s_{SAM,0}$ are computed as follows:

$$s_{SAM}(i) = \sqrt{\frac{(1/n_+)+(1/n_-)}{(n_++n_--2)}} \left[\sum_{j=1}^{n_+}(x_j(i)-\mu_+(i))^2 + \sum_{j=1}^{n_-}(x_j(i)-\mu_-(i))^2 \right],$$

$s_{SAM,0} = 1$.

In the case of the mutual information and Chi-square scores, X_i refers to a random variable representing the ith feature, and can take on any of the values in the set $\{x_i\}$.

References

Bach, F. and Jordan, M. (2001). Kernel independent component analysis. *Tech. Rep. UCB//CSD-01-1166, UC Berkeley.*

Baldi, P. and Long, A.D. (2001). A bayesian framework for the analysis of microarray expression data: regularized t-test and statistical inferences of gene changes. *Bioinformatics*, 17:509–519.

Ball, G. and Hall, D. (1967). A clustering technique for summarizing multivariate data. *Behav. Science*, 12:153–155.

Ben-Dor, A., Bruhn, L., and Friedman, N., et al. (2000). Tissue classification with gene expression profiles. *J. Comp. Biol.*, 7:559–584.

Benjamini, Y. and Hochberg, Y. (1995). Controlling the false discovery rate: A practical and powerful approach to multiple testing. *J. Roy. Stat. Soc. B*, 57:289–300.

Blum, A. and Langley, P. (1997). Selection of relevant features and examples in machine learning. *Art. Intell.*, 97(1-2):245–271.

Breiman, L., Friedman, J.H., Olshen, R.A., and Stone, C.J. (1984). *Classification and regression trees*. Wadsworth International Group, Belmont, CA.

Burges, C.J.C. (1998). A tutorial on support vector machines for pattern recognition. *Data Mining and Knowledge Discovery*, 2(2):121–167.

Chernoff, H. and Lehmann, E.L. (1954). The use of maximum likelihood estimates in chi2 tests for goodness-of-fit. *The Annals of Mathematical Statistics*, 25:576–586.

Cormack, R.M. (1971). A review of classification. *J. Roy. Stat. Soc. A*, 134:321–367.

Cox, T. and Cox, M. (1994). *Multidimensional Scaling*. Chapman & Hall, London.

Dempster, A.P., Laird, N.M., and Rubin, D.B. (1977). Maximum likelihood from incomplete data via the EM algorithm. *J. Roy. Stat. Soc. B*, 34:1–38.

Denham, M.C. (1994). Implementing partial least squares. *Statistics and Computing*.

Dijkstra, T. (1983). Some comments on maximum likelihood and partial least squares methods. *J. Econometrics*, 22:67–90.

Duda, R.O., Hart, P.E., and Stork, D.G. (2000). *Pattern Classification*. Wiley-Interscience Publication.

Eisen, M.B., Spellman, P.T., Brown, P.O., and Botstein, D. (1998). Cluster analysis and display of genome-wide expression patterns. *Proc. Natl. Acad. Sci. USA*, 95(25):14863–14868.

Furey, T.S., Christianini, N., and Duffy, N., et al. (2000). Support vector machine classification and validation of cancer tissue samples using microarray expression data. *Bioinformatics*, 16(10):906–914.

Golub, T.R., Slonim, D.K., and Tamayo, P., et al. (1999). Molecular classification of cancer: Class discovery and class prediction by gene expression monitoring. *Science*, 286:531–537.

Good, P. (1994). *Permutation Tests: A Practical Guide to Resampling Methods for Testing Hypotheses*. Springer.

Gosser, W. S. (1908). The probable error of a mean. *BIOMETRIKA*, 6:1–25.

Guyon, I. and Elisseeff, A. (2003). An introduction to variable and feature selection. *J. Machine Learning Res.*, 3:1157–1182.

Hanley, J. A. and McNeil, B. J. (1982). The meaning and use of the area under a receiver operating characteristic (ROC) curve. *Radiology*, 143(1):29–36.

Hastie, T., Tibshirani, R., and Friedman, J. (2001). *The Elements of Statistical Learning*. Springer, New York/Berlin/Heidelberg.

Hauskrecht, M., Pelikan, R., and Malehorn, D.E., et al. (2005). Feature selection for classification of SELDI-TOF MS proteomic profiles. *Appl. Bioinf.*, 4(4):227–246.

Jolliffe, I. T. (1986). *Principal Component Analysis*. Springer, New York.

Jutten, C. and Herault, J. (1991). Blind separation of sources, part 1: An adaptive algorithm based on neuromimetic architecture. *Signal Process.*, 24(1):1–10.

Kendall, M.G. (1945). The treatment of ties in ranking problems. *Biometrika*, 33(3):239–251.

Kirkpatrick, S., Gelatt, C., and Vecchi, M. (1983). Optimization by simulated annealing. *Science*, 220:671–680.

Kohavi, R. and John, G. (1998). The wrapper approach. In Liu, H. and Motoda, H., editors, *Feature Selection for Knowledge Discovery and Data Mining*, pages 33–50. Kluwer Academic Publishers, Norwell, MA, USA.

Koza, J. (1995). Survey of genetic algorithms and genetic programming. *Proc. Wescon95:E2. Neural-Fuzzy Technologies and Its Applications, IEEE*, pages 589–594.

Krus, D.J. and Fuller, E.A. (1982). Computer-assisted multicrossvalidation in regression analysis. *Educational and Psychological Measurement*, 42:187–193.

Liu, H. and Setiono, R. (1995). Chi2: Feature selection and discretization of numeric attributes. *Proc. 7th IEEE Intl. Conf. Tools with Artificial Intelligence*, page 88.

MacKay, D. (1992). *Bayesian Methods for Adaptive Models*. PhD thesis, California Institute of Technology.

McLachlan, G., Peel, D., and Prado, P. (1997). Clustering via normal mixture models. *Proc. Am. Stat. Assoc.*, pages 98–103.

McQueen, J. (1967). Some methods for classification and analysis of multivariate observations. *Proc. 5th Berkeley Symp. Math. Stat. Prob.*, pages 281–297.

Neal, R. (1998). Assessing relevance determination methods using DELVE. In Bishop, C.M., editor, *Neural Networks and Machine Learning*, pages 28–32. Springer.

Patel, S. and Lyons-Weiler, J. (2004). caGEDA: A web application for the integrated analysis of global gene expression patterns in cancer. *Appl. Bioinf.*, 3(1):49–62.

Pavlidis, P., Weston, J., Cai, J., and Grundy, W.N. (2001). Gene functional classification from heterogeneous data. In *Proc. 5th Ann. Intl. Conf. Comp. Mol. Biol.*, pages 242–248.

Ross, D.T., Scherf, U., and Eisen, M.B., et al. (2000). Systematic variation in gene expression patterns in human cancer cell lines. *Nat. Gen.*, 24:227–235.

Russel, S. and Norvig, P. (1995). *Artificial Intelligence*. Prentice Hall.

Schölkopf, B. and Smola, A. J. (2002). *Learning with Kernels*. MIT Press.

Slonim, D.K., Tamayo, P., Mesirov, J.P., Golub, T.R., and Lander, E.S. (2000). Class prediction and discovery using gene expression data. *Proc. 4th Ann. Intl. Conf. Comp. Mol. Biol.*, pages 263–272.

Speer, N., Spieth, C., and Zell, A. (2005). Spectral clustering gene ontology terms to group genes by function. *Lecture Notes in Bioinformatics*, 3692:001–012.

Storey, J.D. and Tibshirani, R. (2003). SAM thresholding and false discovery rates for detecting differential gene expression in DNA microarrays. In Parmigiani, G., Garrett, E.S., Irizarry, R.A., and Zeger, S.L., editors, *The Analysis of Gene Expression Data: Methods and Software*, pages 272–290. Springer, New York.

Tusher, V.G., Tibshirani, R., and Chu, G. (2001). Significance analysis of microarrays applied to the ionizing radiation response. *Proc. Natl. Acad. Sci. USA*, 98(9):5116–5121.

Tzannes, N.S. and Noonan, J.P. (1973). The mutual information principle and applications. *Information and Control*, 22(1):1–12.

Vapnik, V. (1995). *The Nature of Statistical Learning Theory*. Springer, New York.

Westfall, P.H. and Young, S.S. (1993). *ResamplingBased Multiple Testing: Examples and Methods for P-value Adjustment*. Wiley.

Wilcoxon, F. (1945). Individual comparisons by ranking methods. *Biometrics Bulletin*, 1:80–83.

Xing, E.P., Jordan, M.I., and Karp, R.M. (2001). Feature Selection for High-Dimensional Genomic Microarray Data. *Proc. 18th Intl. Conf. Machine Learning*, pages 601–608.

8

Resampling Strategies for Model Assessment and Selection

Richard Simon

National Cancer Institute, Rockville, MD, USA.
rsimon@mail.nih.gov

8.1 Introduction

The advent of DNA microarrays and proteomics technology has stimulated the development and use of classification algorithms for biomedical studies. In oncology, for example, a common application is predicting response to treatment based on expression profiling of tumor tissue. Such a classifier could be used as an aid in treatment selection for future patients based on the expression profiles of their tumors. In developing such a classifier, it is important to estimate the predictive accuracy that can be expected for future application of the classifier.

There are, in fact, many considerations that may influence the accuracy of predictions made using a classifier developed on a set of data. These include considerations such as assay drift, inter-laboratory variability in assay procedures, tissue handling and preparation, and systematic differences of new cases with regard to unmeasured or unaccounted for prognostic features. The only fully satisfactory way of assessing predictive accuracy of a classifier for independent cases is to classify a set of independent new cases in a manner that reflects all sources of variability to be experienced in broad application of the classifier. Before expending the extensive time and resources necessary for such an "external validation", one needs a more easily obtained estimate of predictive accuracy that can be obtained from the data set used to develop the classifier.

In this chapter, I will discuss the use of resampling methods to obtain "internal estimates" of predictive accuracy of classification models in applications when the number of candidate predictors (e.g., genes) is much greater than the number of cases (e.g., tumor specimens). Much of the conventional wisdom and many of the published claims about resampling methods are invalid for such settings. For clarity, I will focus on binary classifiers, but the considerations described here apply also to classifiers with more than two classes, and, with some technical modifications, to predicting risk groups using time-to-event data.

Key concepts of this chapter include *classifier, true error rate, resubstitution estimate, resampling estimate, bias,* and *variance.* The classifier is the rule for specifying a predicted class as a function of the values of genomic or proteomic measurements. The classifier to be used for subsequent application generally is the one based on the complete set of data in the developmental study. The true error rate is the error rate expected when that classifier is applied to independent data. The re-substitution estimate is the observed error rate when the classifier is applied to the same data set used for its development. The re-substitution estimate is called a biased estimate of the true error rate because the average value of the difference of the two quantities is not zero. A biased underestimate of the true error rate tends to be too small; positive bias means that the estimate tends to be too large. An estimate may be unbiased but have a large variance. That means that, on average, the estimate is not too large or too small, but for any particular set of data the difference between the estimate and the true error rate may be quite large.

8.2 Basic Concepts

8.2.1 Resubstitution Estimate of Prediction Error

Traditionally, statisticians have often used the re-substitution estimate as a measure of predictive accuracy. With the re-substitution estimate, one uses all the data to develop the classifier, and then measures the accuracy of classification on the same set of data used to develop the classifier. The data available for developing the classifier is a set of cases $i = 1, 2, ...n$, with each case i consisting of a vector of predictor variables $\mathbf{x}_i$ and a binary 0 or 1 class label y_i. Let D denote the set of data for all n cases, and let $C(\mathbf{x}_i, D) = \hat{y}_i$ denote the class predicted for a case with vector of variables $\mathbf{x}_i$ when the set of data D was used to develop the classifier. Then the re-substitution estimate of error is the proportion of the n cases for which $y_i \neq \hat{y}_i$.

The re-substitution estimate tends to underestimate the true error rate because the same data is used for developing the classifier and for measuring prediction accuracy (Simon et al., 2003). When a model and its parameters are selected to maximize the fit to a set of data, the degree of fit is not a proper estimate of the degree of discrepancy to be expected with independent data. To measure "prediction", you must separate the data used for developing the classifier from the data used for evaluating performance.

The re-substitution estimate has been frequently used in statistics because the bias is not too great if the number of cases n is large relative to the number of variables p which are measured on the cases and are candidates for inclusion in the classifier. For microarray and proteomic applications, however, the number of candidate variables are generally orders of magnitude greater than the number of cases. In these circumstances, the resubstitution estimate is so biased as to be worthless and completely misleading.

8.2.2 Split-Sample Estimate of Prediction Error

An improved estimate of the true error rate can often be obtained by splitting the data set D into a learning set L and a test set T. For example, the learning set may contain half of the cases and the test set the other half. This is called the *split-sample method*.

A single fully specified classifier is developed on the learning set L. The learning set is not just used to fit the parameters of the classification model, it is used to determine which variables are included in the model and to determine what form of classifier should be used. For example, should the classifier be a linear discriminant, or a neural network with two nodes in a hidden layer, or should it be a nearest neighbor classifier? The learning set is also used to determine all parameters necessary to specify the completely determined binary classifier to be applied to the cases in the test set. I will denote the single fully specified classifier by $C(\mathbf{x}, L)$ to emphasize that only the data in the learning set L can be used in any way for developing the classifier.

Once a single fully specified classifier is developed on the learning set, it is used to classify the cases in the test set. The classifier is not adjusted or calibrated in any way on the test set; it is simply used to classify the cases in the test set and the proportion of cases in the test set incorrectly classified is the estimate of classification error, $\epsilon_{split-sample}$. Let n_T denote the number of test cases. $C(\mathbf{x}_i, L) = \hat{y}_i$ is the predicted class for the ith test case, $\mathbf{x}_i$, made by classifier developed using the learning set. Then

$$\epsilon_{split-sample} = \frac{1}{n_T} \sum_{i=1}^{n} \delta_i, \tag{8.1}$$

with $\delta_i = 1$ if $y_i \neq \hat{y}_i$ and $\delta_i = 0$ if $y_i = \hat{y}_i$.

There are no well established guidelines for what proportion of the data to use for the learning set and what proportion for the test set, or whether the split should be made randomly or in some systematic manner. Often half to two-thirds of the cases are used for the learning set. The split is often made randomly although in multi-center studies a closer emulation of external validation is obtained if one uses samples from some centers for learning and samples from other centers for testing.

What classifier would be used in an external validation study or for classifying future samples? It would probably be the classifier developed using all the data, i.e., $C(\cdot, D)$ rather than the classifier $C(\cdot, L)$ developed on the learning set. The split-sample process is used to provide an estimate of the classifier $C(\cdot, D)$, although the estimate is based on a classifier developed on less data. If the data set D is large, then the classifier based on the full data set may be about as accurate as that based on the learning set. For small data sets, however, the classifier based only on the learning set may be substantially inferior to that based on the full data and the estimate (8.1) of prediction error may tend to be too large (i.e., positively biased). Molinaro et al. (2005) evaluated the split-sample method with 50% or 67% of the data used for the learning set and found that for small data sets the split-sample estimate can be seriously positively biased.

Attempting to avoid the positive bias resulting from having too small a learning set may, however, leave such a small test set that the estimate will have a very large variance. Using the split sample method, the number of test set errors will have a binomial distribution and so the variance of the test set error rate is approximately proportional to the reciprocal of the sample size of the test set (Snedecor and Cochran, 1989).

8.3 Resampling Methods

The difficult tradeoffs between the size of the learning set and test set can be somewhat ameliorated by using a large proportion of the samples for the learning set, but repeating the estimate for various learning-testing partitions and averaging the resulting error estimates. Let $(L_1, T_1), (L_2, T_2), ...(L_m, T_m)$ denote a set of learning-testing partitions of the data D. For each learning set L_i, develop a classifier from scratch. That means that all aspects of classifier development must be repeated for each learning set, including determination of which features are included in the classifier. The full data set cannot be used for any aspect of the development of the classifiers $C(\cdot, L_1), C(\cdot, L_1), ...C(\cdot, L_m)$. Each of these classifiers is applied to the cases not used in its learning set. Let n_{T_j} denote the number of cases in the test set T_j. Then the error estimates are computed as follows:

$$\epsilon_j = \frac{1}{n_{T_j}} \sum_{i=1}^{n_{T_j}} \delta_i, \tag{8.2}$$

with $\delta_i = 1$ if $y_i \neq \hat{y}_i$ and $\delta_i = 0$ if $y_i = \hat{y}_i$. These error estimates are then averaged as follows:

$$\epsilon_{resampling} = \frac{1}{n_T} \sum_{j=1}^{m} \epsilon_j \tag{8.3}$$

The resampling estimate is an estimate of the prediction error for the classifier $C(\cdot, D)$ developed on the full data set. It is based on averaging error estimates for a set of different classifiers, one for each learning set generated by the resampling. Resampling methods can only be used when a completely defined algorithm is used for all aspects of the classifier development process. This is because it is not valid to use information from analysis of the full data set in building a classifier from any learning set, and it is not valid to use information from building a classifier in one learning set for building a classifier in a different learning set. This is a serious limitation for large projects where multiple individuals are involved in the classifier development process and extensive analysis, including biological interpretation, is used in the classifier development process. Consequently, the split-sample method remains a useful approach for many studies.

8.3.1 Leave-One-Out Cross-Validation

A wide variety of different methods have been proposed for constructing the set of partitions $(L_1, T_1), (L_2, T_2), ...(L_m, T_m)$ and leave-one-out cross-validation (LOOCV) is one of the most commonly used methods. With LOOCV, the number of partitions m equals the number of cases, n. Each test set consists of a different singleton set; i.e., $T_i = (\mathbf{x}_i, y_i)$, and each learning set consists of all $n - 1$ cases *not* in the corresponding test set; i.e., $L_i = D \backslash T_i$. Since each learning set contains only one fewer cases than the full data set, the positive bias of the split-sample method is almost eliminated. The high variance of the estimate of prediction error for any single test set is reduced by averaging over all of the n learning-testing partitions. Since any two learning sets have $n - 2$ cases in common, there is a high correlation among the models developed on the different learning sets and a high correlation among the error estimates. Although the n estimates of prediction error for the n test sets are almost completely unbiased, they are positively correlated and this correlation serves to increase the variance of the error estimate (Ambroise and McLachlan, 2002). Molinaro et al. (2005) found that for very small samples (e.g., 40 or fewer cases), the large bias of other methods such as the split sample method is more serious than the increased variance of LOOCV. For larger sample sizes, the bias of the split-sample method decreases, and the computational effort of LOOCV also increases. There are some other alternatives, however, that are discussed below.

Although any two learning sets contain $n - 2$ cases in common, it is still essential to repeat the classifier building process from scratch for each learning set. The studies of Ambroise and McLachlan (2002) and of Simon and Lam (2005) showed that failure to re-select the features independently for each learning set results in hugely biased error estimates. Simon et al. (2003) performed a simulation, for example, in which the distribution of expression levels for all genes were completely unrelated to the class indicators. Since the classes were equally represented, any models developed from the data should have a prediction error of about 0.50. The simulations involved 6 000 genes and 14 cases (7 from each class). They used a linear classifier based on the ten genes that seemed most differentially expressed between the classes. They found that if they used all the data to select the ten most differentially expressed genes, to fit the linear model, and then to measure prediction error, the resulting re-substitution estimate was 0 in over 95% of the simulations. This was not so surprising as the re-substitution estimate was known to be biased. It did emphasize, however, just how biased the re-substitution estimate was for settings where the number of candidate variables was much larger than the number of cases. Simon et al. also evaluated what they called "partial cross-validation". That is, they used all 14 cases to select the ten genes that seemed most differentially expressed between the classes. They then performed LOOCV, re-fitting the linear classifier to each training set, but always using the same ten genes. What they found was that the partial cross-validation error estimate was zero in over 90% of the simulations. So the partial LOOCV estimate is very misleading, almost as biased as the re-substitution estimate.

Leave-one-out cross-validation is sometimes termed the "jackknife" method. The term "jackknife", however, has traditionally referred to estimates of bias or variance based on leave-one-out sub-sampling (Efron and Tibshirani, 1993).

8.3.2 k-fold Cross-Validation

The k-fold cross-validation is a popular alternative to LOOCV. Ten-fold cross validation is a special case of k-fold cross validation. With this approach the cases are randomly partitioned into ten disjoint sets of approximately equal size, denoted $T_1, T_2, ...T_{10}$. As indicated by the notation, each set forms a test set to be used with the corresponding learning set $L_i = D \backslash T_i$ consisting of the remaining 90% of cases not included in the test set T_i. The classifier development process is repeated ten times, and the resulting classifiers are used to classify the cases in their respective test set to produce error estimates $\epsilon_1, \epsilon_2, ...\epsilon_{10}$, which are then averaged as in Equation (8.3). Ten-fold cross-validation requires much less computing than LOOCV when the number of cases is large. It is only slightly more positively biased than LOOCV, since the classifiers are developed on 90% of the full data, and it generally provides error estimates with less variance. Consequently, it is quite popular when there are more than 20 cases. The k-fold cross-validation error estimate can be repeated for different random k-fold partitions and the results averaged. Molinaro et al. (2005), however, found that such averaging had a very limited effect on reducing the variance of the estimate. It can sometimes be useful, however, to design the k-fold partition to be used in a manner that is balanced with regard to a variable correlated with class label.

8.3.3 Monte Carlo Cross-Validation

Monte Carlo cross-validation, also known as *repeated random subsampling*, is a generalization of the split-sample method. It repeats the process of splitting the data into a learning set and a test set, with the model developed on the learning set and the error rate evaluated on the test set. With Monte Carlo cross-validation, the test set estimates are averaged over the learning-testing random splits. For each learning-testing partition, each case appears in either the learning set or the test set, but not in both. Because the learning-testing splits are random, the cases appear in the test sets a variable number of times. That is, one case may appear in m learning sets and in no test set, whereas another case may appear in all test sets, but in none of the learning sets. With a large number of repetitions, m, however, such extreme differences would be unusual. In k-fold cross-validation, the test sets are disjoint, whereas in Monte Carlo cross-validation, the test sets usually overlap. Molinaro et al. (2005) found that Monte Carlo cross-validation does not reduce the bias of the split-sample method, but it can substantially reduce the variance of the split-sample error estimate. They also reported that ten random learning-testing splits were sufficient to realize most of the achievable reduction in variance.

8.3.4 Bootstrap Resampling

Several versions of the *bootstrap* have been proposed for estimation of the true prediction error. Each bootstrap learning set is composed of n cases selected with replacement from the full set of n cases. Consequently, cases may appear in a learning set multiple times. The bootstrap learning set differs from the learning sets constructed by the other methods in that it contains n cases, albeit not n distinct cases. If L_i denotes the ith bootstrap learning set, the corresponding test set is the set of cases not included in L_i, that is, $T_i = D \backslash L_i$. With the *leave-one-out bootstrap*, the prediction error is estimated as the average of the error rates of the classifiers $C(\cdot, L_i)$ evaluated on the test sets (Efron and Tibshirani, 1993). Although the learning sets contain n cases, since they are not all distinct cases, the leave-one-out bootstrap tends to over-estimate the true error rate.

8.3.4.1 The .632 Bootstrap

To attempt to correct for the positive bias of the leave-one-out bootstrap, Efron and Tibshirani (1993) developed two alternative bootstrap methods. These methods were based on weighted averages of the estimate of the leave-one-out bootstrap and the re-substitution estimate. For the .632 bootstrap, the weight of the leave-one-out bootstrap estimate is 0.632 and the weight for the learning set is 0.368 (Efron and Tibshirani, 1993).

A bootstrapped learning set is generated by randomly sampling n cases uniformly with replacement. Each case has a probability of $p = 1 - n^{-1}$ of *not* being selected. If the data set is sampled n times, then the probability that a case is not selected for the bootstrap sample is $p = (1 - n^{-1})^n \approx e^{-1} = 0.368$ (for large n). Hence, the expected number of distinct cases in the bootstrap learning set is $0.632 \times n$. The error estimate on the test set will be pessimistic, since the model is trained on 63.2% of the data only. Therefore, the error estimate on the learning set is combined with the error estimate test set with different weights. If b bootstrapped data sets are generated, then the estimate for the classification error is $\epsilon_{.632}$:

$$\epsilon_{.632} = .632\epsilon_{loobs} + (1 - 0.632)\epsilon_{resub} \qquad (8.4)$$

$$\epsilon_{loobs} = \frac{1}{b}\sum_{i=1}^{b}\epsilon_{T_i} \qquad (8.5)$$

$$\epsilon_{resub} = \frac{1}{b}\sum_{i=1}^{b}\epsilon_{L_i}, \qquad (8.6)$$

where ϵ_{T_i} is the observed error rate on the ith test set and ϵ_{L_i} is the observed error rate on the corresponding i^{th} learning set

The .632 bootstrap estimate can be very downward biased, however, for high-dimensional data (Molinaro et al., 2005). For example, consider the situation where the expression data is uninformative for predicting the class variable. In that case the true prediction error is 0.5 for equally represented classes, and the leave-one-out bootstrap is unbiased. The leave-one-out bootstrap is unbiased because there is really no penalty for developing classifiers based on a reduced number of distinct cases in the learning set, since no classifier that performs better than the flip of a coin is possible. Even in this situation, however, the re substitution estimate can be close to zero as shown by (Simon et al., 2003). Since the .632 bootstrap estimate is a weighted average of the leave-one-out bootstrap estimate and the re-substitution estimate, the result is also downward biased.

8.3.4.2 The .632+ Bootstrap

Efron and Tibshirani (1997) developed the .632+ bootstrap to attempt to improve on the bias of the .632 bootstrap. With the .632+ bootstrap, the weight for the leave-one-out bootstrap is not a fixed value but adjusted based on an estimate of the degree to which the data is overfit.

$$\epsilon_{.632+} = w \cdot \epsilon_{loobs} + (1 - w)\epsilon_{resub} \qquad (8.7)$$

$$w = \frac{.632}{1 - 0.368R} \qquad (8.8)$$

$$R = \frac{\epsilon_{loobs} - \epsilon_{resub}}{\epsilon_{random} - \epsilon_{resub}} \qquad (8.9)$$

The quantity R is supposed to provide an estimate of the degree of overfitting. ϵ_{random} is an estimate of the expected error when the class labels are independent of all prediction variables. Molinaro et al. (2005) found that the .632+ bootstrap performed well except for high-dimensional data in cases where the classes were well separated. In those cases, the .632+ estimate could be much greater than the true value.

8.4 Resampling for Model Selection and Optimizing Tuning Parameters

Many classification algorithms contain one or more tuning parameters that must be specified before a completely specified binary classifier is obtained. For example, linear support vector machine classifiers with inner product kernels contain two parameters that must be determined. One is equivalent to a ridge regression shrinkage parameter and the other represents the relative cost of misclassification of cases in the two classes. The *shrunken centroid classification algorithm* of Tibshirani et al. (2002) contains a parameter which determines the amount of shrinkage, and ultimately the number of variables used for classification. We might represent these classifiers by $C(\mathbf{x}, \lambda_i, D)$ indicating that one has a set of classifiers, indexed by the values of the tuning parameters λ_i. Often resampling methods, such as those described above, are used to obtain estimates of the prediction error for a set of such classifiers. Let us denote the true prediction error for the classifier $C(\mathbf{x}, \lambda, D)$ by $\tau(\lambda_i, D)$ and the resampling estimator by $\epsilon(\lambda_i, D)$. It is reasonable to select a vector of tuning parameters to be used for future classification as that which minimizes the estimated error rate; that is:

$$\lambda^* = \arg\min_{\lambda_i}\{\epsilon(\lambda_i, D)\} \tag{8.10}$$

It is important to recognize, however, that the resampling estimate of prediction error at λ^* is not an unbiased estimate of the prediction error that can be expected for the selected classifier $C(\mathbf{x}, \lambda^*, D)$. That is, $\epsilon(\lambda^*, D)$ is not an unbiased estimator of $\tau(\lambda^*, D)$. A bias is created in the process of minimizing over the set of tuning parameters. Varma and Simon (2006) have studied this bias and indicated that its size depends on the number of classifiers minimized over and the variability of the resampling estimates. The process of optimizing over values of the tuning parameters, however, should be viewed as an integral part of the algorithm for determining a completely specified classifier. As emphasized above for the case of selecting genes, it is the entire algorithm for determining a completely specified classifier which needs to be embedded in the resampling algorithm. In order to obtain a proper estimate of $\tau(\lambda^*, D)$, one should re-select the "optimal" tuning parameter vector for each resampled learning set. Since the selection of an optimal tuning parameter involves the use of a resampling estimator, one needs to implement a *two-fold nested resampling* in order to accomplish this. The *outer loop* of resampling is for the purpose of estimating the prediction error of a completely specified classifier. The *inner loop* of resampling is for the purpose of defining a completely specified classifier by optimizing the tuning parameters (Varma and Simon, 2006). In the inner loop, the learning set is split into a *training set* and a *validation set*, as illustrated in Figure 8.1.

The resampling produces multiple classifiers, constructed on multiple training sets and tested on their associated validation sets. This process is,

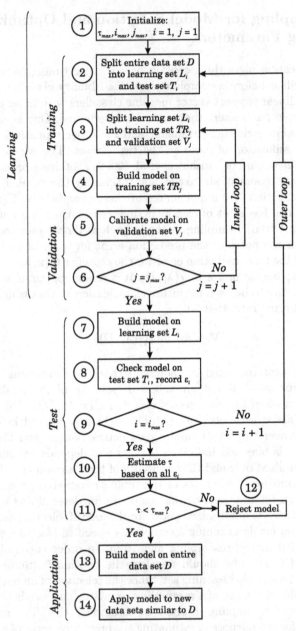

Fig. 8.1. Basic modeling process involving learning (training and validation), testing, and application.

however, just a statistical procedure to estimate the prediction accuracy of the model fitted to the full data set. Data analysts are sometimes tempted to use the similarities and differences among the classifiers developed on the resampled training sets for purposes of model selection. This happens particularly often with regard to feature selection. That is, a defined feature selection is used for classifier development in each training set. The analyst is often surprised at how much the sets of selected features differ among the training sets and they are tempted to use for the final classifier only those features that were consistently selected in most training sets. That, however, represents a completely new algorithm for classifier development. When one is using resampling for estimating the true error rate of a classifier, then all aspects of the algorithm used for developing the classifier must be repeated in each resampled learning set. If part of the algorithm involves determining which variables to select based on their frequency of selection in resampling the data, then the resampling process must be performed for variable selection for each learning set of the resampling loop used for error estimations.

8.4.1 Estimating Statistical Significance of Classification Error Rates

In addition to providing a fair estimate of prediction accuracy for a classifier, it is usually important to test the null hypothesis that the prediction accuracy is no better than one could have obtained by chance. One can address this question by obtaining the distribution of resampled error estimates for data sets in which there is no relationship between the candidate prediction variables and the class labels. Such data sets can be generated by randomly permuting the assignment of class labels to cases, keeping fixed the number of cases for each class. For each permutation D^*, the resampling procedure is completely repeated resulting in a new classifier $C(\mathbf{x}, D^*)$ and a new error estimate $\epsilon_{resampling}(D^*)$. This is done for either all possible permutations or for a large number of random permutations, and the area in the left tail of this distribution corresponding to error rates no greater than that obtained with the un-permuted labels $\epsilon_{resampling}(D)$ represents the significance level of the test of the null hypothesis. This approach was first described by Radmacher et al. (2002). Michiels et al. (2005) proposed an alternative test based on whether a prediction interval for the resampled prediction error estimate excluded 0.5, but it is less powerful because its inefficient use of the data in training-validation splits.

8.4.2 Comparison to Classifiers Based on Standard Prognostic Variables

Another question that often arises is whether the genomic or proteomic data enables better predictions than were already possible using standard clinical or histopathological variables. Some investigators attempt to address this by using resampling to obtain a class prediction for each case, and using these predictions in a classification model that involves the predictions and the other variables. For example, let $\hat{y}_i$ denote the predicted class for case i based on a classifier developed using a learning set not containing case i. One might then study a logistic regression model containing these genomic or proteomic predictions as well as other variables. The problem is, however, that although the predictions were based on learning sets not containing the cases predicted, the predictions are generally not statistically independent and so the resulting logistic inference is not valid. These problems have been noted by Tibshirani and Efron (2002).

Kattan (2003, 2004) points out that prognostic models should be judged based on their ability to predict, not based on issues of the significance of regression coefficients for some variables after adjustment for other variables.

8.5 Comparison of Resampling Strategies

Molinaro et al. (2005) conducted a detailed study of different resampling methods for estimating prediction error in the context of high-dimensional data where the number of candidate predictors is much larger than the number of cases. They evaluated the methods with several different classification methods including diagonal linear discriminant analysis, linear discriminant analysis, nearest neighbor classification, and CART classification trees. They examined the resampling methods and classification algorithms in the context of data sets of different sizes, different numbers of informative features and different signal strengths (class separation relative to noise). For small data sets (e.g., 40 cases), they found that the split sample methods with one-half to two thirds of cases used for training produced very biased over-estimates of prediction error. Monte Carlo repeats of the split samples had no effect on the strong bias. For such data sets they found that LOOCV, 10-fold cross-validation and the .632+ bootstrap generally had the least bias and smallest mean-square error. The .632+ bootstrap was more biased than LOOCV and 10-fold cross-validation, however, in small data sets with strong signals. The differences among resampling methods decreased as the sample size increased.

8.6 Tools and Resources

Investigators and data analysts need to exercise caution in use of packaged software because much of it is does not use complete resampling methods essential for high-dimensional data. *BRB-ArrayTools* software incorporates all of the resampling methods described in this chapter for a wide variety of classifier types. The software is available without cost for non-commercial applications (Simon and Lam, 2005).

8.7 Conclusions

Resampling-based estimates of prediction error are not substitutes for fully independent validation studies, but are essential for helping to identify when independent validation studies are warranted.

The re-substitution estimate commonly employed is highly biased for genomics and proteomics applications and should not be used. The split-sample method is useful for large studies in which many analyses and many individuals are involved in model development. The test data should not be used in any way until a single, completely specified classifier is developed and agreed to by the collaborators. At that time the classifier should be applied to classify the cases in the test set, with no calibration or adjustment of the model. The prediction accuracy of the classifier is scored on the test cases.

For smaller projects the split-sample approach represents an inefficient use of the data compared to resampling methods. Resampling methods, however, require that the classifier development process be algorithmic and repeatable in its entirety on multiple learning sets.

Feature selection is an integral part of classifier development and must be repeated from scratch for each learning set when resampling is used.

If resampling is used to optimize tuning parameters of a classifier, then that resampling should be viewed as an integral part of classifier development and the entire algorithm should be embedded in an outer resampling loop used to estimate the prediction accuracy of the resulting classifier.

Resampling methods provide an estimate of the prediction error that can be expected for future use of the classifier fit to the entire data set. It is the classifier developed on the full data set that is taken forward. The resampling produces multiple classifiers, constructed on the multiple training sets and tested on their associated validation sets. This process is, however, just a statistical method to estimate the prediction accuracy of the model fitted to the full data set, the intermediate models should not be used in the definition of the classifier to be taken forward.

For small data sets, leave-one-out cross-validation or 10-fold cross validation provide almost unbiased estimates of prediction error for a variety of classifiers and a wide variety of signal strengths. The mean-square error of these methods compares favorably to those of sample-splitting based methods that provide too few cases in the training set and thereby result in over-estimates of prediction accuracy. For larger data sets, 10-fold cross-validation is preferable to LOOCV with regard to computational effort and variance. The standard leave-one-out bootstrap and .632 bootstrap can be quite biased for small data sets. The .632+ bootstrap is generally competitive with LOOCV and 10-fold cross-validation for small data sets except those with very strong signals.

References

Ambroise, C. and McLachlan, G.J. (2002). Selection bias in gene extraction on th basis of microarray gene expression data. *Proc. Natl. Acad. Sci. USA*, 98:6562–6566.

Efron, B. and Tibshirani, R. (1993). *An Introduction to the Bootstrap*. Chapman & Hall.

Efron, B. and Tibshirani, R.J. (1997). Improvements on cross-validation. The .632+ bootstrap method. *J. Am. Stat. Assoc.*, 92:548–560.

Kattan, M.W. (2003). Judging new markers by their ability to improve predictive accuracy. *J. Natl. Cancer Inst.*, 95(9):634–635.

Kattan, M.W. (2004). Evaluating a new marker's predictive contribution. *Clin. Cancer Res.*, 10:822–824.

Michiels, S., Koscielny, S., and Hill, C. (2005). Prediction of cancer outcome with microarrays: A multiple random validation strategy. *The Lancet*, 365:488–492.

Molinaro, A.M., Simon, R., and Pfeiffer, R.M (2005). Prediction error estimation: A comparison of resampling methods. *Bioinformatics*, 21(15):3301–3307.

Radmacher, M.D., McShane, L.M., and Simon, R. (2002). A paradigm for class prediction using gene expression profiles. *J. Comp. Bio.*, 9(3):505–511.

Simon, R. and Lam, A.P. (2005). BRB-ArrayTools Users Guide (version 3.3). Technical Report 28, Biometric Research Branch, National Cancer Institute, Bethesda, MD, USA.

Simon, R., Radmacher, M.D., Dobbin, K., and McShane, L.M. (2003). Pitfalls in the analysis of DNA microarray data: Class prediction methods. *J. Natl. Cancer Inst.*, 95:14–18.

Snedecor, G.W. and Cochran, W.G. (1989). *Statistical Methods*. Iowa State University Press, Ames Iowa, USA.

Tibshirani, R., Hastie, T., Narasimhan, B., and Chu, G. (2002). Diagnosis of multiple cancer types by shrunken centroids of gene expression. *Proc. Natl. Acad. Sci. USA*, 99(10):6567–6572.

Tibshirani, R.J. and Efron, B. (2002). Pre-validation and inference in microarrays. *Stat. Appl. Gen. Mol. Biol.*, 1(1).

Varma, S. and Simon, R. (2006). Bias in error estimation when using cross-validation for model selection. *BMC Bioinformatics*, 7:91.

Classification of Genomic and Proteomic Data Using Support Vector Machines

Peter Johansson and Markus Ringnér

Computational Biology and Biological Physics Group, Department of Theoretical Physics, Lund University, Sweden.
peter@thep.lu.se, markus@thep.lu.se

9.1 Introduction

Supervised learning methods are used when one wants to construct a classifier. To use such a method, one has to know the correct classification of at least some samples, which are used to train the classifier. Once a classifier has been trained it can be used to predict the class of unknown samples. Supervised learning methods have been used numerous times in genomic applications and we will only provide some examples here. Different subtypes of cancers such as leukemia (Golub et al., 1999) and small round blue cell tumors (Khan et al., 2001) have been predicted based on their gene expression profiles obtained with microarrays. Microarray data has also been used in the construction of classifiers for the prediction of outcome of patients, such as whether a breast tumor is likely to give rise to a distant metastasis (van't Veer et al., 2002) or whether a medulloblastoma patient is likely to have a favorable clinical outcome (Pomeroy et al., 2002). Proteomic patterns in serum have been used to identify ovarian cancer (Petricoin et al., 2002a) and prostate cancer (Adam et al., 2002; Petricoin et al., 2002b).

In this chapter, we will give an example of how supervised learning methods can be applied to high-dimensional genomic and proteomic data. As a case study, we will use support vector machines (SVMs) as classifiers to identify prostate cancer based on mass spectral serum profiles.

9.2 Basic Concepts

In supervised learning the aim is often to construct a rule to classify samples in pre-defined classes. The rule is constructed by learning from learning samples. The correct class assignments are known for the learning data and used in the construction of the rule. The number of samples needed to construct a classification rule is data set and classification method dependent. In our experience at least ten samples in each class are needed for classification of genomic and proteomic data. Once the rule is constructed it can be applied to classify unknown samples. Each sample typically is a vector of real values and each value often corresponds to a measurement of one feature of the sample. Classification rules can be either explicit or implicit. An example of a classification method with an implicit rule is the nearest neighbor classifier: A test sample is classified to belong to the same class as the learning sample to which it is most similar. Decision trees are a classification method that use explicit rules, for example, if the value of a specific feature of the sample is positive, the sample is predicted to belong to one class, if not, it is predicted to belong to another class. In the case of classifying tumor samples based on genomic or proteomic data, each sample could, for example, be a vector of gene expression levels from a microarray experiment, volumes from spots on a 2-dimensional gel, or intensities for different mass-over-charge (m/z) values from a mass spectrum. Classes could, for example, be cancer patients and healthy individuals, respectively. In these cases, data sets have very many features and it is often beneficial to separate construction of the classifier in two parts: Feature selection and classifier rule construction. Moreover, once a classifier is constructed its predictive performance needs to be evaluated. In this section, we will discuss a supervised learning method: SVM, how feature selection can be combined with SVMs, and a methodology for obtaining estimates of the predictive performance of the classifier.

9.2.1 Support Vector Machines

Suppose we have a set of samples where each sample belongs to one of two pre-defined classes, and we have measured two values for each sample, for example, the expression levels of two proteins (Figure 9.1). Two classes of two-dimensional samples are considered linearly separable if a line can be constructed such that all samples of one class lie on one side of the line and all samples of the other class lie on the other side. This line serves as a decision boundary between the two classes. For higher-dimensional data, a linear decision boundary is a hyperplane that separates the classes. If classes are separable by a hyperplane, there are most likely many possible hyperplanes that separate the classes (Figure 9.1a). SVMs are based on this concept of decision boundaries. SVMs are designed to find the hyperplane with the largest distance to the closest points from the two classes, the *maximal margin hyperplane* (Figure 9.1c). Once this hyperplane has been found for a set of learning samples, the class of additional test samples can be predicted based on which side of the hyperplane they appear.

For many classification problems the classes cannot be separated by a hyperplane; they are not linearly separable and a non-linear decision surface may be useful. Classifiers that use a non-linear decision surface are called *non-linear classifiers*. SVMs address such classification problems by mapping the data from the original input space into a feature space in which a linear separator can be found. This mapping does not need to be explicitly speci-

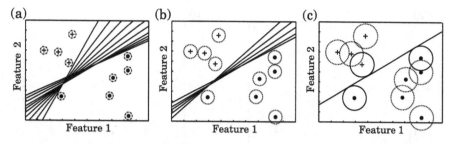

Fig. 9.1. Finding the separating hyperplane with the maximal margin for the linearly separable case. A data set consisting of 10 samples with 2 features each is classified. Each sample belongs to one of two different classes (denoted by + and •, respectively). A linear decision boundary is a hyperplane separating the two classes. In the case of two features such a hyperplane is a line. Many hyperplanes exist that perfectly separate the samples from the two classes (a). Suppose we draw circles with the same diameter around each sample and that the separating hyperplanes are not allowed to intersect the circles. It is then obvious that with an increasing diameter of the circles, the number of allowed hyperplanes decreases (b). The diameter is increased until only one hyperplane exist (c). This hyperplane is completely defined by the points encircled by solid circles and these points are called *support vectors*. Intuitively, this final hyperplane seems more appropriate as a decision boundary because it maximizes the margin between the two classes. SVMs are designed to find this maximal margin hyperplane.

fied. Instead a user of SVMs needs to select a so-called *kernel function*, which can be viewed as a distance between samples in feature space (Figure 9.2). The linear decision surface in feature space may correspond to a non-linear separator in the original input space. To avoid over-fitting to data, avoid sensitivity to outlier samples, or handle problems that are not linearly separable in feature space, SVMs with soft-margins can be used. For such SVMs the strict constraint to have perfect separation between the classes is softened. A parameter denoted C is introduced to tolerate errors. The larger C is, the harder errors are penalized. The limit of C being infinity corresponds to the maximal margin case for which no errors are tolerated.

For microarray and proteomic data, for which the number of features is much larger than the number of samples, it is typically possible to find a linear classifier that perfectly separates the samples. It is our experience for high-dimensional data that SVMs with a linear kernel and the C parameter set to infinity results in classifiers with better predictive performance than SVMs for which one tries to optimize kernel selection and C parameter value. An SVM with this choice of parameters is called a *linear maximal margin classifier*.

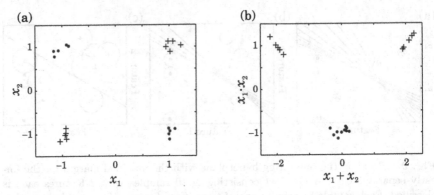

Fig. 9.2. Mapping the XOR problem into a feature space in which it is linearly separable. A data set consisting of 20 samples with two features each and belonging to two different classes (denoted by $+$ and $\bullet$, respectively) is shown. (a) The two classes are not linearly separable in the original input space (x_1, x_2). (b) The input space can be mapped to the feature space $(x_1 + x_2, x_1 x_2)$, where the two classes are linearly separable. The kernel for this mapping is $K(\mathbf{a}, \mathbf{b}) = (a_1 + a_2)(b_1 + b_2) + a_1 a_2 b_1 b_2$.

9.2.2 Feature Selection

The amount of information contained in a mass spectrum, or the number of genes measured on a microarray, results in very many features for each sample. When the number of features is much larger than the number of samples, it represents a challenge for many supervised learning methods. This problem is often referred to as the "curse of dimensionality". To address this problem, feature selection techniques are used to reduce the dimensionality of the data to improve subsequent classification. Feature selection methods can be divided into two broad categories: *Wrapper methods* and *filter methods*.

Wrapper methods evaluate feature relevance within the context of the classification rule. For example, by first constructing a classification rule using all features, and then analyzing the classification rule to identify the features most important for the rule, followed by constructing a new classification rule using only these important features. A wrapper method used with SVMs is *recursive feature elimination* (Guyon et al., 2002).

Filter methods select features based on a separation criterion unrelated to the classification rule. For example, the standard two-sample t-test could be used as a filter criterion for two class classification problems to identify and rank features that discriminate between the two classes. Another commonly used filter selection criteria used to rank features is the signal-to-noise ratio (S2N ratio) (Golub et al., 1999). Here each feature is ranked based on S2N $= |\mu_+ - \mu_-|/(\sigma_+ - \sigma_-)$, where $\mu_\pm$ and $\sigma_\pm$ are the average and standard deviation, respectively, of the values for the feature in the two classes $+$ and $-$ (see Equation(7.1), Chapter 7). A classification rule is then constructed using only the features that have the largest S2N.

Filter methods have been shown to provide classification performances that are comparable to or outperform other selection methods for both microarray and mass spectrometry data (Wessels et al., 2005; Levner, 2005). In our experience t-test and S2N perform similarly and the choice of the filter criterion is not crucial.

9.2.3 Evaluating Predictive Performance

To evaluate the predictive performance of a classifier, one needs a test set that is independent of all aspects of classifier construction. If the number of samples investigated is relatively small, one often resorts to cross-validation. In n-fold cross-validation, the samples are randomly split into n groups of which one is set aside as a test set and the remaining groups are a learning set used to calibrate a classifier. The procedure is then repeated with each of the n groups used as a test set. Finally, the samples can again be randomly split into n groups and the whole procedure repeated many times. These test sets would provide a reliable estimate of the true predictive performance, if there were no choices in classifier construction. An example of such a case is if one, prior to any data analysis, decides to use an SVM with linear kernel, C set to infinity, and all features (no feature selection). However, suppose one only wants to use the features that provide the best prediction results, then the test set is no longer independent because it has been used to optimize the number of features to use in the classification rule. To circumvent this use of the test set, the learning samples can be used to optimize the prediction performance of the classifier in an internal procedure of cross-validation. This internal n-fold cross-validation is identical to the cross-validation used for generating test sets, except that one group of samples is used to optimize the performance of the classifier (validation set) and the remaining samples are used to train the classifier (training set). Hence, these cross-validation samples will provide an overly optimistic estimate of the predictive performance (Ambroise and McLachlan, 2002). Once all the choices required to construct a classifier has been made in the internal cross-validation, the performance can be evaluated on the samples set aside in the external cross-validation loop. A schematic picture of this procedure is given in Figure 9.3. The number of folds used in the external and internal cross-validations can be different. To assess if a predictive performance achieved by SVMs is significant random permutation tests can be used. In these tests the predictive performance is compared to results for SVMs applied to the same data but with the class labels randomly permuted (Pavey et al., 2004).

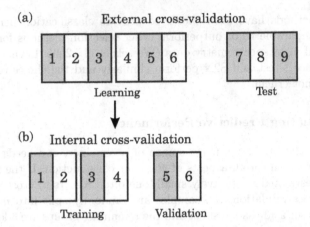

Fig. 9.3. One fold in an (a) external and (b) internal 3-fold cross-validation procedure. In total, the data set comprises nine samples, each represented by a number. The external cross-validation is used to estimate the predictive performance of the classifier and the internal cross-validation is used to optimize the choices made when constructing the classifier. The samples belong to two classes, gray and white. For both the internal and external cross-validation, the folds are stratified to approximate the same class distributions in each fold as in the complete data set.

9.3 Advantages and Disadvantages

9.3.1 Advantages

- SVMs perform non-linear classification by mapping data into a space where linear methods can be applied. In this way, non-linear classification problems can be solved relatively fast computationally.

9.3.2 Disadvantages

- SVMs may be too sophisticated for many genomic and proteomic classification problems (Somorjai et al., 2003). Occam's razor principle tells us to prefer the simplest method and often linear SVMs give the best classification performance, in which case alternative linear classification methods may be more easily applied.

9.4 Caveats and Pitfalls

The two most important aspects of classification of genomic and proteomic data are not directly related to the choice of classification method. First, high-quality data needs to be obtained, in which biologically relevant features are not confounded by experimental flaws. Second, a proper methodology to evaluate the classification performance needs to be implemented to avoid overly optimistic estimates of predictive performances, or more importantly, to avoid finding a classification signal when there is none (Simon et al., 2003). When estimating the true predictive performance using a test data set, it is crucial to use a procedure, in which the test data is not used to select features, to construct the classification rule, or even to select the number of features to use in the classifier. In the case study, we will see how the violation of these requirements influences classification results.

9.5 Alternatives

Nearest centroid classifiers provide an alternative to SVMs that are simple to implement and have been used successfully for many genomic applications (van't Veer et al., 2002; Tibshirani et al., 2002; Wessels et al., 2005). For this type of classifiers there exists available software tailored for genomic and proteomic data. Therefore, we describe how a simple version is implemented. First, the arithmetic mean for each feature is calculated using only samples within each class. In this way a prototype pattern for each class called a *centroid* is obtained. Second, one defines a distance measure between samples and centroids, and the classes of additional test samples are predicted by calculating to which centroid they are nearest.

There are many supervised learning methods that can be applied to genomic and proteomic data, including linear discriminant analysis, classification trees, and nearest neighbor classifiers (Dudoit et al., 2002), as well as artificial neural networks (Khan et al., 2001).

9.6 Case Study: Classification of Mass Spectral Serum Profiles Using Support Vector Machines

As a case study, we applied SVMs to a public data set of mass spectral serum profiles from prostate cancer patients and healthy individuals (Petricoin et al., 2002b) to see how well the disease status of these individuals could be predicted.

9.6.1 Data Set

The data set consists of 322 samples: 63 samples from individuals with no evidence of disease, 190 samples from individuals with benign prostate hyperplasia, 26 samples from individuals with prostate cancer and PSA levels 4 through 10, and 43 samples from individuals with prostate cancer and PSA levels above 10. For our case study, we followed previous analysis (Levner, 2005) and combined the two first groups into a healthy class containing 253 samples, and the latter two groups into a disease class containing 69 samples.

For each individual, a mass spectral profile of a serum sample has been generated using surface-enhanced laser desorption ionization time-of-flight (SELDI-TOF) mass spectrometry (Hutchens and Yip, 1993; Issaq et al., 2002). In this method, a serum sample taken from a patient is applied to the surface of a protein-binding chip. The chip binds a subset of the proteins in the serum. A laser is used to irradiate the chip resulting in the proteins being released as charged ions. The time-of-flights of the charged ions are measured providing an m/z value for each ion. Each sample will in this way give rise to a spectrum of intensity as a function of m/z; a proteomic signature of the serum sample. The data set used in our case study consists of spectra with intensities for 15 154 m/z values. Hence, the number of features (15 154) is much larger than the number of samples (322).

9.6.2 Analysis Strategies

We used linear SVMs with C set to infinity (linear maximal margin classifiers). The predictive performance was evaluated using 5 times repeated 3-fold external cross-validation, which resulted in a total of 15 test sets, each with one-third of 322 samples. The cross-validation was stratified to approximate the same class distributions in each fold as in the complete data set. We used S2N to rank features and classifiers using different numbers of top-ranked features (n_f) were evaluated. A set of classifiers was constructed, in which each classifier was trained using $1.5 \times n_f$ more top-ranked features than the previous classifier; the first classifier used the top-ranked feature ($n_f = 1$) and the final classifier used all features. The performance of classifiers was evaluated using the balanced accuracy (BACC). Given two classes, 1 and 2 with N_1 and N_2 samples, respectively, we denote samples known to belong to class 1 as true positives (TP) if they are predicted to belong to class 1, and false negatives (FN) if they are predicted to belong to class 2. Correspondingly, samples known to belong to class 2 are true negatives (TN) if they are predicted to belong to class 2 and false negative (FN) otherwise. BACC is the average of the sensitivity and the specificity, in other words, it is the average of the fractions of correctly classified samples for each of the two classes:

$$\mathrm{BACC} = \frac{1}{2}\left(\frac{\mathrm{TP}}{\mathrm{TP}+\mathrm{FN}} + \frac{\mathrm{TN}}{\mathrm{TN}+\mathrm{FP}}\right) = \frac{1}{2}\left(\frac{\mathrm{TP}}{N_1} + \frac{\mathrm{TN}}{N_2}\right) \qquad (9.1)$$

An advantage of BACC is that a simple majority classifier that predicts all samples into the most abundant class will obtain a BACC of 50% even though it is expected to obtain overall classification accuracies, $(\mathrm{TP}+\mathrm{TN})/(N_1+N_2)$, higher than 50%.

We used four different strategies to construct SVMs : Two strategies in which the test sets were independent of all aspects of SVM construction, and two strategies exemplifying how overly optimistic estimates of the predictive power can be obtained (Figure 9.4).

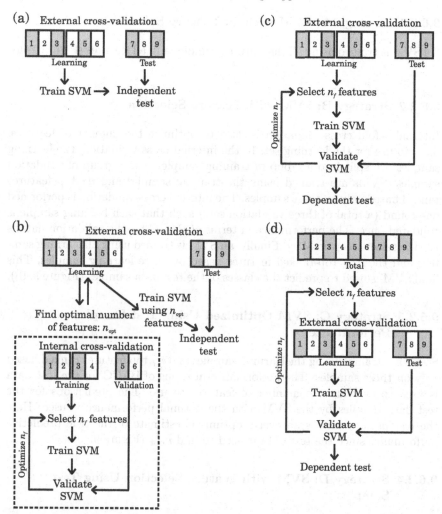

Fig. 9.4. The four strategies used to construct SVMs. In strategies A and B the test samples are not used for training SVMs, for feature selection, or for optimizing the number of features to use. Hence, a good estimate of the true predictive performance may be obtained using strategies A or B. In strategy C the test samples are used to select the number of features to use. In strategy D the test samples are used both to select features and to optimize the number of features to use. Hence, an overly optimistic estimate of the true predictive performance is obtained using strategies C or D.

9.6.2.1 Strategy A: SVM without Feature Selection

An SVM is trained using all the learning samples and all features (Figure 9.4a).

9.6.2.2 Strategy B: SVM with Feature Selection

Internal 3-fold cross-validation is used to optimize the number of features, n_{opt}, to use for the learning set. In the internal cross-validation, the learning samples are split into a group of training samples and a group of validation samples. SVMs are trained using the training samples and n_f top features ranked based only on these samples. The internal cross-validation is performed one round (a total of three validation sets) such that each learning sample is validated once. The performance in terms of BACC for the validation data is used to select an optimal n_f. Finally, an SVM is trained using all learning samples and the n_{opt} top-ranked features based on these learning samples. This final SVM is used to predict the classes of the test data samples (Figure 9.4b).

9.6.2.3 Strategy C: SVM Optimized Using Test Samples Performance

SVMs are trained using the learning samples and n_f top features ranked based only on these samples. The performance in terms of BACC for the test data is used to evaluate the number of features to use. The predictions for the test data samples by the SVM with the optimal performance is used. Here the test performance is an overly optimistic estimate of the true predictive performance since the test data is used to find n_{opt} (Figure 9.4c).

9.6.2.4 Strategy D: SVM with Feature Selection Using Test Samples

SVMs are trained using the learning samples and n_f top features ranked based on all samples. The performance in terms of BACC for the test data is used to evaluate the number of features to use. The predictions for the test data samples by the SVM with the optimal performance is used. Here the test performance is an overly optimistic estimate of the true predictive performance since the test data is used both to find n_{opt} and to rank features (Figure 9.4d).

9.6.3 Results

The results in terms of BACC for the four different strategies are summarized in Table 9.1.

Table 9.1. Predictive performance of SVMs

Strategy[a]	BACC(%)[b]	
	Mean	Std
A: SVM without feature selection	88.7	3.6
B: SVM with feature selection	91.1	3.3
C: SVM optimized using test sample performance	94.6	2.6
D: SVM with feature selection using test samples	94.9	1.8

[a]See Analysis Strategies in Section 9.6.
[b]Balanced accuracy: average of sensitivity and specificity.

9.7 Lessons Learned

We have shown an example of how SVMs are capable of predicting with high accuracy whether mass spectral serum profiles belong to a healthy or a prostate cancer class. High balanced accuracy (88.7%) was obtained without any feature selection, yet a simple filter selection method improved the predictive performance and a BACC of 91.1% was obtained. This BACC is competitive with the best performance obtained for this data set in a study of different feature selection methods (Levner, 2005). In the context of cross-validation it is difficult to evaluate if a method is significantly better than another method because different test sets have samples in common (Berrar et al., 2006).

There are many pitfalls when evaluating the predictive performance of classifiers. By using the test data simply to select the number of features to use by the classification rule, the BACC increased to 94.6%. This performance is an overly optimistic estimate of the true predictive performance not likely to be achieved for independent test data. Similarly, overly optimistic results were obtained when the test data was used to rank features prior to feature selection. It is important to realize that overly optimistic evaluations may lead to incorrect conclusions for classes which cannot be classified (Ambroise and McLachlan, 2002; Simon et al., 2003).

9.8 List of Tools and Resources

There are several publicly available implementations of SVMs and a comprehensive list is available at http://www.kernel-machines.org/software.html. For example, there is an implementation in C called SVMLight (http://svmlight.joachims.org/) and an implementation called LIBSVM with interfaces to it for many programming languages (http://www.csie.ntu.edu.tw/~cjlin/libsvm). In the case of microarray data analysis, SVMs are available as a part of the TM4 microarray software suit (http://www.tm4.org/).

Publicly available implementations of nearest centroid classifiers include ClaNC (Dabney, 2006);

http://students.washington.edu/adabney/clanc/) and PAM (Tibshirani et al., 2002); http://www-stat.stanford.edu/~tibs/PAM/), both implemented for the R package (http://www.r-project.org).

9.9 Conclusions

SVMs can be used to classify high-dimensional data such as microarray or proteomic data. Often the simplest SVMs called maximal margin linear SVMs are able to obtain high accuracy predictions. For many applications it is important to evaluate classifiers based on their predictive performance on test data. In this evaluation, it is important to implement test procedures that do not lead to overly optimistic results. We have used mass spectral serum profiles of prostate cancer patients and healthy individuals as a case study to exemplify how the predictive performance of a classifier can be estimated. We conclude that SVMs predict the samples in our case study with high balanced accuracy.

9.10 Mathematical Details

We will consider the simplest version of a support vector machine, the so-called linear maximal margin classifier for classification of data points in two classes. This classifier only works for data points which are linearly separable. Two classes of two-dimensional samples are considered linearly separable if a line can be constructed such that all cases of one class lie on one side of the line and all cases of the other class lie on the other side. For a more detailed description of SVMs, there are many books available for the interested reader (Vapnik, 1995; Burges, 1998; Cristianini and Shawe-Taylor, 2001). Consider a linearly separable data set $\{(\mathbf{x}_i, y_i)\}$, where $\mathbf{x}_i$ are the input values for the ith data point and y_i is the corresponding class $\{-1, 1\}$. The assumption, that the data set is linearly separable, means that there exists a hyperplane that separates the data points of the two classes without intersecting the classes. This hyperplane serves as a decision surface, and we can write:

$$\mathbf{w}'\mathbf{x}_i + b \geq 0 \; i : y_i = +1$$
$$\mathbf{w}'\mathbf{x}_i + b \leq 0 \; i : y_i = -1 \,,$$

where the hyperplane is defined by $\mathbf{w}$ and b, and $\mathbf{w}'\mathbf{x}+b$ is the output function. The distance from the hyperplane to the closest point is called the *margin* (denoted by γ). The underlying idea of the maximal margin classifier is that, in order to have a good classifier, we want the margin to be maximized. We notice there is a free choice of scaling: Rescaling $(\mathbf{w}, b)$ to $(\lambda\mathbf{w}, \lambda b)$ does not change the classification given by the output function. The scale is to set such that

$$\mathbf{w}'\mathbf{x}^+ + b = +1$$
$$\mathbf{w}'\mathbf{x}^- + b = -1 \, , \tag{9.2}$$

where $\mathbf{x}^+$ ($\mathbf{x}^-$) is the closest data point on the positive (negative) side of the hyperplane. Now it is straightforward to compute the margin

$$\gamma = \frac{1}{2}\left(\frac{\mathbf{w}'\mathbf{x}^+ + b}{\|\mathbf{w}\|} - \frac{\mathbf{w}'\mathbf{x}^- + b}{\|\mathbf{w}\|}\right) = \frac{1}{\|\mathbf{w}\|} \, .$$

Hence, when the scale is set such that Equation (9.2) is fulfilled then maximizing the margin is equivalent to minimizing the norm of the weight vector, $\|\mathbf{w}\|$. This can be formulated as a quadratic ($\mathbf{w}'\mathbf{w}$) problem with inequality constraints ($y_i(\mathbf{w}'\mathbf{x}_i + b) \geq 1$):

$$\text{min:} \ \frac{1}{2}\mathbf{w}'\mathbf{w} \quad \text{subject to:} \ y_i(\mathbf{w}'\mathbf{x}_i + b) \geq 1 \text{ for all } y_i.$$

The Lagrangian for this quadratic problem is

$$L(w, b, \alpha) = \frac{1}{2}\mathbf{w}'\mathbf{w} - \sum_{i=1}^{M}\alpha_i[y_i(\mathbf{w}'\mathbf{x}_i + b) - 1] \, , \tag{9.3}$$

where $\alpha_i \geq 0$ are the Lagrange multipliers and M is the number of data points. Differentiating with respect to w and b and setting the partial derivatives to zero give

$$\frac{\partial L}{\partial \mathbf{w}} = 0 \Rightarrow \mathbf{w} = \sum_{i=1}^{M}\alpha_i y_i \mathbf{x}_i \tag{9.4}$$

$$\frac{\partial L}{\partial b} = 0 \Rightarrow \sum_{i=1}^{M}y_i \alpha_i = 0 \, , \tag{9.5}$$

and putting this into Equation (9.3) gives the Wolfe dual

$$Q(\alpha) = \frac{1}{2}\sum_{i=1}^{M}\alpha_i y_i \mathbf{x}_i' \sum_{j=1}^{M}\alpha_j y_j \mathbf{x}_j - \sum_{i=1}^{M}\alpha_i\left[y_i\left(\sum_{j=1}^{M}\alpha_j y_j \mathbf{x}_j' \mathbf{x}_i + b\right) - 1\right]$$

$$= -\frac{1}{2}\alpha' H\alpha + \sum_{i=1}^{M}\alpha_i \, , \tag{9.6}$$

where the elements of the matrix H are given by $H_{ij} = y_i y_j \mathbf{x}_i' \mathbf{x}_j$. The original problem is transformed into a dual problem. Finding the dual vector, α, that maximizes Q and fulfills the constraint in Equation (9.5) is equivalent to solving the original problem (Kuhn and Tucker, 1951).

The output of the classifier for a data point $\mathbf{x}$ can be expressed in terms of the dual vector

$$o(\mathbf{x}) = \mathbf{w}'\mathbf{x} + b = \sum_{i=1}^{M} \alpha_i y_i \mathbf{x}_i'\mathbf{x} + b \ , \tag{9.7}$$

and we note that the weight vector is never needed explicitly. The variable b is set such that the conditions in Equation (9.2) are fulfilled. The binary classification of a data point $\mathbf{x}$ is sign($o(\mathbf{x})$).

The hyperplane that separates the data points is completely defined by a subset of the data points and these data points are called support vectors (Figure 9.1). Each α_i corresponds to a learning data point $\mathbf{x}_i$ and α_i is zero for all data points except the support vectors. Hence the sum in Equation (9.7) only gets contributions from the support vectors.

In summary, there are three steps to build and use a maximal margin classifier. First, the class labels of the learning data points should be set to $+1$ or -1. Second, the quadratic function in Equation (9.6) is minimized subject to the constraint in Equation (9.5) and all $\alpha_i \geq 0$. Because the matrix H in Equation (9.6) is positive definite there are no local minima and there is a unique solution to the minimization problem. Finally, this solution is used to classify data points in a validation/test set using Equation (9.7).

We conclude by briefly outlining how to extend this linear classifier to non-linear SVMs. The basic idea underlying non-linear SVMs is to map data points into a feature space in which an optimal hyperplane can be found as outlined for the maximal margin classifier. This hyperplane may then correspond to a non-linear separator in the original space of data points. A key observation is that in our construction of the maximal margin classifier we only use the scalar product between data points, $\mathbf{x}_i'\mathbf{x}_j$. If we have a non-linear mapping, $\mathbf{x} \mapsto \varphi(\mathbf{x})$, of data points into feature space, the scalar product between two vectors in feature space, called a kernel function, is

$$K(\mathbf{x}_i, \mathbf{x}_j) = \varphi(\mathbf{x}_i)'\varphi(\mathbf{x}_j) = \sum_l \varphi_l(\mathbf{x}_i)\varphi_l(\mathbf{x}_j) \ .$$

In SVMs, the scalar product between data points used to calculate both H and $o(\mathbf{x})$ is replaced by the kernel K. Hence, to build and use an SVM, the mapping into feature space itself can be ignored and only a kernel function needs to be defined.

References

Adam, B.L., Qu, Y., Davis, J.W., Ward, M.D., Clements, M.A., Cazares, L.H., Semmes, O.J., Schellhammer, P.F., Yasui, Y., Feng, Z., and G.L. Wright, Jr. (2002). Serum protein fingerprinting coupled with a pattern-matching algorithm distinguishes prostate cancer from benign prostate hyperplasia and healthy men. *Cancer Res.*, 62(13):3609–3614.

Ambroise, C. and McLachlan, G.J. (2002). Selection bias in gene extraction on the basis of microarray gene-expression data. *Proc. Natl. Acad. Sci. USA*, 99(10):6562–6566.

Berrar, D., Bradbury, I., and Dubitzky, W. (2006). Avoiding model selection bias in small-sample genomic datasets. *Bioinformatics*, 22(10):1245–1250.

Burges, C. (1998). A tutorial on support vector machines for pattern recognition. *Data Mining and Knowledge Discovery*, 2(2):121–167.

Cristianini, N. and Shawe-Taylor, J. (2001). *An Introduction to Support Vector Machines and other Kernel-based Learning Methods*. Cambridge University Press.

Dabney, A.R. (2006). ClaNC: point-and-click software for classifying microarrays to nearest centroids. *Bioinformatics*, 22(1):122–123.

Dudoit, S., Fridlyand, J., and Speed, T.P. (2002). Comparison of discrimination methods for the classification of tumors using gene expression data. *J. Am. Stat. Assoc.*, 97:77–87.

Golub, T.R., Slonim, D.K., Tamayo, P., Huard, C., Gaasenbeek, M., Mesirov, J.P., Coller, H., Loh, M.L., Downing, J.R., Caligiuri, M.A., Bloomfield, C.D., and Lander, E.S. (1999). Molecular classification of cancer class discovery and class prediction by gene expression monitoring. *Science*, 286(5439):531–537.

Guyon, I., Weston, J., Barnhill, S., and Vapnik, V. (2002). Gene Selection for Cancer Classification using Support Vector Machines. *Machine Learning*, 46:389–422.

Hutchens, T.W. and Yip, T.T. (1993). New desorption strategies for the mass spectrometric analysis of macromolecules. *Rapid. Commun. Mass Spectrom.*, 7:576–580.

Issaq, H.J., Veenstra, T.D., Conrads, T.P., and Felschow, D. (2002). The SELDI-TOF MS approach to proteomics: Protein profiling and biomarker identification. *Biochem. Biophys. Res. Commun.*, 292(3):587–592.

Khan, J., Wei, J.S., Ringnér, M., Saal, L.H., Ladanyi, M., Westermann, F., Berthold, F., Schwab, M., Antonescu, C.R., Peterson, C., and Meltzer, P.S. (2001). Classification and diagnostic prediction of cancers using gene expression profiling and artificial neural networks. *Nat. Med.*, 7(6):673–679.

Kuhn, H.W. and Tucker, A.W. (1951). Nonlinear programming. *Proc. 2nd Berkeley Symp. Mathematical Statistics and Probability*, pages 481–492.

Levner, I. (2005). Feature selection and nearest centroid classification for protein mass spectrometry. *BMC Bioinformatics*, 6(1):68.

Pavey, S., Johansson, P., and Packer, L., et al. (2004). Microarray expression profiling in melanoma reveals a BRAF mutation signature. *Oncogene*, 23(23):4060–4067.

Petricoin, E.F., Ardekani, A.M., and Hitt, B.A., et al. (2002a). Use of proteomic patterns in serum to identify ovarian cancer. *Lancet*, 359(9306):572–577.

Petricoin, E.F., Ornstein, D.K., and Paweletz, C.P., et al. (2002b). Serum proteomic patterns for detection of prostate cancer. *J Natl Cancer Inst*, 94(20):1576–1578.

Pomeroy, S.L., Tamayo, P., and Gaasenbeek, M., et al. (2002). Prediction of central nervous system embryonal tumour outcome based on gene expression. *Nature*, 415(6870):436–442.

Simon, R., Radmacher, M.D., Dobbin, K., and McShane, L.M. (2003). Pitfalls in the use of DNA microarray data for diagnostic and prognostic classification. *J. Natl. Cancer Inst.*, 95(1):14–18.

Somorjai, R.L., Dolenko, B., and Baumgartner, R. (2003). Class prediction and discovery using gene microarray and proteomics mass spectroscopy data: curses, caveats, cautions. *Bioinformatics*, 19(12):1484–1491.

Tibshirani, R., Hastie, T., Narasimhan, B., and Chu, G. (2002). Diagnosis of multiple cancer types by shrunken centroids of gene expression. *Proc. Natl. Acad. Sci. USA*, 99(10):6567–6572.

van't Veer, L., Dai, H., and van de Vijver, M.J., et al. (2002). Gene expression profiling predicts clinical outcome of breast cancer. *Nature*, 415(6871):530–536.

Vapnik, V. (1995). *The Nature of Statistical Learning Theory*. Springer Verlag.

Wessels, L., Lodewyk, F.A., Reinders, M.J.T., Hart, A.A.M., Veenman, C.J., Dai, H., He, Y.D., and van't Veer, L.J. (2005). A protocol for building and evaluating predictors of disease state based on microarray data. *Bioinformatics*, 21(19):3755–3762.

Networks in Cell Biology

Carlos Rodríguez-Caso[1] and Ricard V. Solé[1,2]

[1] ICREA-Complex Systems Lab, Universitat Pompeu Fabra (GRIB), Dr Aiguader 80, 08003 Barcelona, Spain.
carlos.rodriguez@upf.edu

[2] Santa Fe Institute, 1399 Hyde Park Road, New Mexico 87501, USA.
ricard.sole@upf.edu

10.1 Introduction

Both natural and artificial systems can be understood as the interaction of a given set of elements. Interactions lead to global behavior often beyond the simple sum of the properties of each element. Interactions create most behaviors around us: A meeting between two people, file transfers among computers, predator-prey dynamics, cell responses, complex protein formation or DNA-protein binding. From these interactions large-scale systems emerge as a mesh of relations: Society, Internet, food webs, organisms, tissues or cells. Such organizations cannot be reduced to individual properties and a global view is required.

Network (or *graph*) *theory* tackles the study of large-scale systems considering the relations among elements as an abstraction (a graph) where the elements are nodes and the relations are links (edges). This approach has been used for the analysis of natural (Milgram, 1967; Ferrer and Solé, 2001; Solé and Montoya, 2001) and artificial networks (Albert et al., 1999; Vázquez et al., 2002). Real networks are sparse and they are likely to exhibit *scale-freeness, small-world pattern* and *modularity*. As Figure 10.1 shows, networks exhibit different patterns of organization depending on the properties that they have. Understanding and measuring these differences is at the heart of complex networks theory.

The view of molecular biology has been modified due to the necessity of managing the massive data acquired by means of high-throughput techniques. Such an approach reveals that genes, proteins and metabolites are interacting in a network. This perception can be tackled suitably in the context of graph theory. As we will see in this chapter, this theory provides visualization algorithms and topological measures to analyze networks at different levels; from the small groups of elements to the whole graph. Moreover, it gives an integrative framework to acquire a global perspective beyond the traditional reductionistic views of molecular biology.

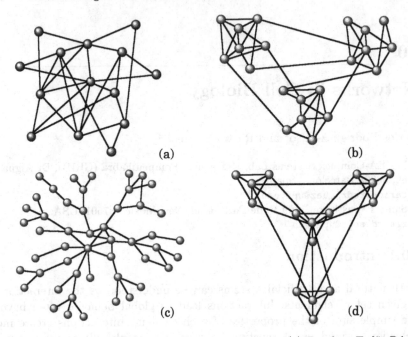

Fig. 10.1. Different examples of network organization. (a) Random Erdös-Rényi graph; (b) random modular graph; (c) scale-free; (d) hierarchical, modular scale-free network (Barabasi and Oltvai, 2004).

The main aim of this chapter is to present to the molecular biologist and bioinformatician those relevant concepts of graph theory for the analysis of cellular networks.

Cellular network is the term commonly used to refer to the current interacting molecular sets within cells (Albert, 2005; Barabasi and Oltvai, 2004). It includes mainly *protein-protein interaction, metabolism, gene expression* and *signal transduction pathways*. All of them are different subsets of a single large-scale cellular network, since they are eventually cross-linked.

10.1.1 Protein Networks

Proteomes, interactomes and protein maps refer to networks of proteins interacting by physical contact. Proteins are the nodes and physical interaction among them are the links in the graph. Large-scale studies have explored the proteome structure in viruses (McCraith et al., 2000), yeast (Uetz et al., 2000; Ito et al., 2001; Ptacek et al., 2005), the worm *Caenorhabditis elegans* (Walhout et al., 2000; Li et al., 2004), *Helicobacter pylori* (Rain et al., 2001), *Drosophila melanogaster* (Giot et al., 2003) and more recently in humans (Rual et al., 2005; Stelzl et al., 2005). Protein map elucidation is obtained mainly by two large-scale experimental approaches, namely, the yeast two-hybrid (Y2H) (Uetz and Hughes, 2000) and the tandem affinity purification (TAP) followed by mass spectroscopy (Gavin et al., 2002). Such information is collected in annotated databases. MIPS (http://mips.gsf.de/), DIP (http://dip.doe-mbi.ucla.edu/) and BIND (http://www.bind.ca/Action) are the main commonly used for the acquisition of current protein maps.

10.1.2 Metabolic Networks

Metabolism is the best described cellular network so far. However, a global topological view of metabolism was not available until recently (Jeong et al., 2000; Ouzounis and Karp, 2000). Since two types of elements participate in metabolism (metabolites and reactions), it allows building networks in different ways. One way is considering the substrate graph, where each metabolite is a node that will be linked with those metabolites participating in the same reaction (Wagner and Fell, 2001; Tanaka, 2005). Alternatively, a reaction graph (Tanaka, 2005) is made by considering reactions as nodes and metabolites as links. The information required to build these graphs is also collected in databases. A compilation of metabolic pathways can be found in KEGG database (http://www.genome.jp/kegg/).

As it occurs with protein maps, it has been observed that very few nodes have many links, whereas most nodes have only a few. The highly connected nodes are referred to as *hubs*, which in metabolic networks correspond to acetyl-CoA and pyruvate, among others (Ravasz et al., 2002).

10.1.3 Transcriptional Regulation Maps

The transcriptional network provides the map of regulatory relations among genes through transcription factors (TFs). TFs are gene products regulating the expression of genes by the interaction with their promoter regions (Shen-Orr et al., 2002). Transcriptional graphs are commonly depicted indicating the arrow of the interaction and the resulting graphs are known as *directed graphs*. Most gene targets for transcription factors are not involved in gene regulation and only receive links, e.g., metabolic enzymes, or structural components of the cell. By contrast, transcription factor genes are also regulated, so they are both *sources* and *sinks* of links.

Gene regulatory networks are built from genome-wide expression analysis designed to reverse-engineer the architecture (Wyrick and Young, 2002; Lee et al., 2002). Additionally, genome-wide location analysis is used to know the transcription factors bound to a given promoter region (Lee et al., 2002). Examples include *E. coli* (Shen-Orr et al., 2002; Salgado et al., 2004) and the yeast regulatory network (Guelzim et al., 2002; Lee et al., 2002).

10.1.4 Signal Transduction Pathways

These networks depict those processes allowing cells integrating responses to external stimuli. They are a combination of metabolic reactions and protein interactions that trigger specific changes in gene expression. Protein modifications such as phosphorylation, acetylation and ubiquitination, among others, lead to conformational changes allowing ligand-protein recognition and functional protein complexes assembling. Kinases and phosphatases are at the basis of the best described pathways. Bibliographic sources provide the current information to reconstruct these kind of networks (Ma'ayan et al., 2005). Additionally, several databases compile this information such as the *Kinbase* (http://kinase.com/) or the commercial database *Protein lounge* (http://www.proteinlounge.com/).

10.2 Basic Concepts

We will present a basic theoretical framework oriented to describe and analyze cellular networks. We will start with providing those descriptors to define, in a topological way, an element within a network. Second, we will provide global descriptors to define a network and finally we will present some more accurate algorithms to uncover the web structure. As we will see for the case study, these methods can be used for the analysis in two ways, *i*) identifying elements that might play a relevant biological function and *ii*) understanding – from an evolutionary point of view – what processes might be important in shaping network organization.

10.2.1 Graph Definition

A *graph* (or network) G is defined by a set of N vertices (or nodes) $V = \{v_1, v_2, ..., v_N\}$ and a set of L edges (or links), $E = \{e_1, e_2, ..., e_L\}$, linking the nodes. Two nodes are linked when they satisfy a given condition, such as two metabolites participating in the same reaction in a metabolic network. The graph definition does not imply that all nodes must be connected in a single component. A *connected component* in a graph is formed by a set of elements so that there is at least one path connecting any two of them.

Graphs are *undirected* when the interaction between nodes is mutual and equal, as in the protein maps. Otherwise, the web is *directed*, as for gene regulatory networks (Shen-Orr et al., 2002) and signal transduction pathways (Ma'ayan et al., 2005). Additionally, graphs can also be *weighted* when links have values according to a certain property. This is the case for gene regulatory networks, where weights indicate the strength and direction of regulatory interactions.

Although graphs are usually represented as a plot of nodes and connecting edges, they can also be defined by means of the so-called *adjacency matrix*, i.e., an array A of $N \times N$ elements a_{ij}, where $a_{ij} = 1$ if v_i links to v_j and zero otherwise. A is symmetric for undirected graphs, but not for the directed ones. For weighted nets a matrix W can be introduced, where w_{ij} indicates the strength and type of the link. The network can also be described using a list of pairs of connected nodes (edge-list), which has some computational advantages. Figure 10.2 summarizes the different ways of representing a graph.

Network theory makes use of a very simple class of networks, so-called *Erdös-Rényi* (ER) graphs (Erdös and Rényi, 1960), to which real networks are compared with. An ER network is defined as a set of nodes linking to each other with a certain probability P. In other words, P defines for each a_{ij} in the adjacency matrix the probability of finding $a_{ij} = 1$. In ER models it is possible to relate P and different network descriptors (see below).

10.2.2 Node Attributes

Here we summarize those measures required to describe individual nodes of a network. They allow to identify elements by their topological properties.

The *degree* (k_i) of a node v_i is defined as the number of edges of this node. The degree is also termed as *connectivity*. From the adjacency matrix, we have $k_i = \sum_{j=1}^{N} a_{ij}$. For directed graphs, we distinguish between incoming and outgoing links. Thus, we specify the degree of a node in its *indegree* k_{in} and *outdegree* k_{out}. See examples of k values in Figure 10.2.

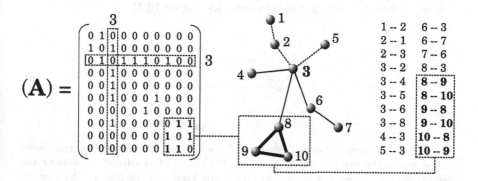

Fig. 10.2. Different ways of graph representation. Left: Adjacency matrix. Center: Drawn graph. Right: List of pairs (edge list). The triangle motif (in bold) is indicated for the three representations. A *pathway* among v_1 and v_5 is depicted by dashed links. Some examples of k, C and b values: for v_3, $k_3 = 5$, $C_3 = 0$, $b_3 = 0.69$; v_8, $k_8 = 3$, $C_8 = 0.33$, $b_8 = 0.36$; v_{10}, $k_{10} = 2$, $C_{10} = 1$, $b_{10} = 0$.

The *clustering coefficient* C_i is a local measure quantifying the likelihood that neighboring nodes of v_i are connected with each other. It is calculated by dividing the number of neighbors of v_i that are actually connected among them, n, with all possible combinations excluding self-links, i.e., $k_i(k_i - 1)$. Formally, we have:

$$C_i = \frac{2n}{k_i(k_i - 1)} \tag{10.1}$$

See examples of C values in Figure 10.2.

The *betweenness centrality* b_m for a node v_m is the fraction of *shortest pathways* Γ for each pair of nodes (v_i, v_j) also containing v_m, that is

$$b_m = \sum_{i \neq j} \frac{\Gamma(i, m, j)}{\Gamma(i, j)} \tag{10.2}$$

The ratio $\Gamma(i, m, j)/\Gamma(i, j)$ indicates how crucial v_m is in relating v_i and v_j. We introduce the term *pathway* (or simply *path*) as the string of nodes relating v_i and v_j (see graph and values for b in Figure 10.2). This concept is similar to the metabolic pathway describing a set of coupled reactions from one metabolite to another. The shortest path connecting v_i and v_j is the one where the lowest number of nodes are involved to connect them.

Such topological descriptors are useful to identify particular nodes in the network. Under this viewpoint, such particularities can be mapped into relevant topological properties. For instance, high k_i for a node might relate to a relevant role, since many other nodes interact with it. Alternatively, high b_i can also indicate a relevant role since it tells us that many nodes are efficiently connected through it. It is noteworthy that, b_i usually scales with degree, although this is not always true (see Figure 10.3).

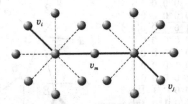

Fig. 10.3. Relation between degree and betweenness. The two-star graph shows a case where the two hubs support a high level of shortest pathways, however the central node v_m shows the highest b of the graph keeping a low degree. The *shortest path* connecting v_i and v_j through v_m is indicated by solid lines.

10.2.3 Graph Attributes

For a network of size N, several useful global measures can be defined, each one providing very different, but complementary, sources of information. *Average degree*, defined as $\langle k \rangle = 2L/N$, indicates how sparse a graph is. For ER graphs, $\langle k \rangle = PN$. Most real networks are sparse, i.e., $\langle k \rangle \ll N$.

Average clustering $\langle C_i \rangle = (1/N) \sum_i C_i$ provides a measure of local organization. High $\langle C_i \rangle$ indicates that neighbors of a node are likely to be linked between them. It actually gives the probability of finding triangles. For ER graphs, $\langle C_i \rangle = \langle k \rangle / N$.

Average path length (ℓ) indicates the average length of the shortest pathways separating each node pair. If d_{min} is the length of the shortest path connecting nodes v_i and v_j, then ℓ is defined as:

$$\ell = \frac{2}{N(N-1)} \sum_{i>j} d_{min}(v_i, v_j) \qquad (10.3)$$

For ER graphs ℓ follows the expression $\ell_{ER} = logN/log\langle k \rangle$. When a network G_Ω fulfills the conditions $\ell_\Omega \simeq \ell_{ER}$ but $\langle C \rangle_\Omega \gg \langle C \rangle_{ER}$ then it is said that G_Ω exhibits a *small-world* (SW) *pattern*. Figure 10.4 shows a graph with a SW pattern (also called *small-world graph*) between two extreme networks: An ordered lattice (a) and a pure random network (c). The SW graph (b) retains the high clustering displayed by the lattice web (a), but it also has very small ℓ, as expected from ER in (c). It is obtained by rewiring a small number of links (Watts and Strogatz, 1998). These networks keep their local order (high C) but also allow a very efficient communication (low ℓ).

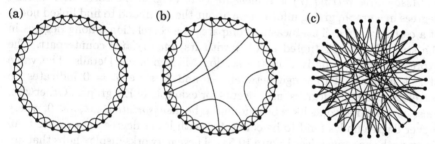

Fig. 10.4. Randomization process from regular lattice (a) to a pure random network (c); (b) represents a small-world graph where local relations are still conserved (as in (a)) but the average path length ℓ is close to the one in (c).

Another measure is the *degree distribution* p_k. It indicates the probability of a node having k links and thus $\sum_k p_k = 1$. It illustrates one of the most striking differences seen in real networks as compared to ER ones. As defined, ER networks follow a binomial distribution, which for large N can be approximated by a Poisson distribution, indicating that the majority of the nodes are close to the average degree. Most real networks exhibit a degree distribution following a power-law decay, $p_k \sim k^{-\gamma}$. Here, γ is a positive parameter that for real networks is usually in the range $2 < \gamma < 3$ (Albert and Barabasi, 2002). Networks fulfilling this property are known as *scale-free*.

Scale-free graphs have a p_k with a maximum at $k = 1$ (thus most elements have a single link) and rapidly decay at higher k values. However, the tail of the distribution is very long and thus nodes with a very high degree are possible. In contrast, ER graphs predict that very high k is exceedingly rare and unlikely to be observed at all. SF distributions have no humps and have extremely large standard deviations, which means that no confidence can be placed in a prediction of the number of links of any node sampled at random (Albert and Barabasi, 2002). Typically, real networks exhibit a mixed distribution, that is, a power-law with a sharp exponential cut-off determined by k_c in the expression $p_k \sim (k)^{-\gamma} e^{-\frac{k}{k_c}}$ indicating that arbitrarily high degrees are not allowed (Amaral et al., 2000).

Clustering distribution (C_k) represents C_i against k. ER and pure scale-free webs (graphs (a) and (c) in Figure 10.1, respectively) do not exhibit any dependency between C_i and k. By contrast, in so-called *hierarchical networks* (see graph (d) in Figure 10.1), C_k decays as the inverse of the degree ($C_k \sim k^{-1}$) (Barabasi and Oltvai, 2004). This type of network exhibits modularity (nodes are preferentially linked inside clusters or modules). A *module* can be defined as a set of nodes in a connected component which tend to be more connected among them than with the rest of the network.

The *betweenness distribution* (b_k) is defined in a similar way, i.e., it represents b_i against k. It has been shown that real networks, such as the Internet (Vázquez et al., 2002), follow a power-law b_k.

Assortative mixing (r) is a measure that weights the correlation among degrees in a graph, giving information about the likelihood to find linked nodes of a certain degree. This measure compares the correlation among degrees in the studied network (noted as G_Ω) with its *uncorrelated* counterpart. The expression for r is derived in the section Mathematical Details. The value of r is normalized to range between -1 and 1. Here, $r = 0$ indicates no correlations among degrees, as it occurs for example in ER graphs. Otherwise, most complex networks have been found to be *disassortative*, i.e., $r < 0$, where higher degree nodes tend to be connected with lower degree ones rather than nodes with the same k (see Figure 10.5a). These networks display hubs that are not directly connected among them. It has been suggested that this situation confers network robustness (Maslov and Sneppen, 2002). When $r > 0$, nodes with the same degree tend to be linked among them (see Figure 10.5b) and the graph is called *assortative*.

Correlation profiles (Maslov and Sneppen, 2002) are related to r, but defined for each pair of degree, i.e., (k_0, k_1). The studied network G_Ω is compared with a set of random graphs with the same degree distribution p_k, known as *randomized graphs* G_r. They are obtained through a rewiring process where, taking two edges without any vertices in common, we exchange their starting vertices. By iterating this process at least twice the number of edges, a reasonably randomized graph G_r preserving p_k is produced. Denoting by $P_\Omega(k_0, k_1)$ the actual probability in G_Ω of finding an edge linking two nodes with degree k_0 and k_1 and by $P_r(k_0, k_1)$ its randomized counterpart, we define the correlation profile Z as $Z(k_0, k_1) = (P_\Omega(k_0, k_1) - P_r(k_0, k_1))/\sigma_r(k_0, k_1)$.

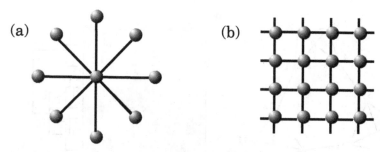

Fig. 10.5. Illustration of *assortativenness*. (a) A star graph, showing a correlation among highly connected nodes with poorly connected ones ($r < 0$). (b) A lattice where all nodes have $k = 4$. It is the extreme case where nodes with the same degree tend to be linked among them ($r > 0$).

It measures the difference between the correlations of G_Ω and G_r, normalized by the standard deviation of $P_r(k_0, k_1)$, indicated by $\sigma_r(k_0, k_1)$.

The Z values are calculated for each pair (k_0, k_1). If $Z = 0$, there is no difference among G_Ω and G_r; $Z > 0$ corresponds to an over-representation of connections in G_Ω among (k_0,k_1) pairs compared with G_r. Whereas under-representation is given by $Z < 0$. Z is depicted as a 2D array of (k_0,k_1) pairs.

Topological overlap analysis is one of the possible algorithms used for detecting modularity. The method arranges nodes depending on the neighbors they share, in other words, two nodes will be closer in the arrangement depending on how many common neighbors they have. Afterwards, they are drawn in a 2D symmetric array where the strength of the relation between each pair is displayed using a color gradient (Ravasz et al., 2002). This algorithm allows building a dendrogram that captures hierarchical relations between nodes. Figure 10.6 shows an example for a simple modular graph.

The *k-scaffold graph analysis* (Rodriguez-Caso et al., 2005) allows us to obtain a well-defined subgraph containing all hubs, and their interaction partners. One pair of connected nodes is conserved, in the so-called *k-scaffold graph*, if the degree of at least one of them is larger than a predefined cut-off k_c. It is noteworthy that the *k*-scaffold not only reveals hubs but also their immediate neighbors. This allows to keep the connections among two hubs through intermediary nodes.

Network motifs are small and repeated patterns of connections among a few nodes. Motifs usually do not exceed 4-5 nodes. We evaluate all the combinations of connections that can be implemented using these small sets. Network motif analysis is based on the dissection of a network into small subgraphs of three to five interacting nodes (Milo et al., 2002). Real networks exhibit an overabundance of certain motifs compared to the randomized ones. Although such deviations have been suggested to relate with functional traits (Milo et al., 2002) they might actually reflect the rules of duplication and divergence driving genome evolution (Solé and Valverde, 2006).

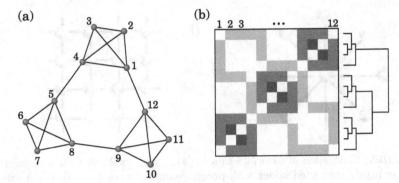

Fig. 10.6. Topological overlap analysis for a simple graph with three modules. Topological overlap matrix (b) depicts the arranged nodes by their overlapping defined by the graph (a). A grey gradient shows the relationship for each pair of nodes, where darker shades indicate the highest overlapping.

10.3 Caveats and Pitfalls

Here, we summarize what a topological perspective can offer as well as its limitations to the study of large-scale networks

- Graph theory is an adequate approach for large-scale systems. It can provide a comprehensive representation of the overall pattern of molecular interactions obtained from experimental approaches.
- Graph theory gives a suitable framework to model large-scale systems. Evolutionary models of proteome evolution based on gene duplication and deletion has been proposed suggesting that these processes give similar topological properties to the observed proteins maps (Pastor-Satorras et al., 2003).
- Network analysis can be used to identify candidates with potential biological relevance by observing their topological features. This is the case for the degree as it is shown in the Case Study section.
- The network definition (in its topological form) is an abstraction requiring a great simplification and thus a loss of information. For example, a protein complex formation in protein maps can depend on a sequential assembling that is not considered in current protein map definitions.
- Current information about different molecular networks is far from being complete. For instance, it has been shown that protein maps coverage are not sufficient to obtain significant information about the degree distribution for the entire proteome (Han et al., 2005).
- Distinct molecular networks are partly embedded inside a very large, multilayered network, involving metabolism, protein interactions and gene regulation. In general, such networks, are studied separately, but the crosslinking among them is somewhat ignored.

10.4 Case Study: Topological Analysis of the Human Transcription Factor Interaction Network

Transcription factors (TFs) are an essential subset of interacting proteins, since they are responsible for the control of gene expression. They interact with DNA regions and tend to form transcriptional regulatory complexes. The final effect of one of these complexes will be determined by its transcription factor composition. The number of transcription factors varies among organisms, but it appears to be linked to their complexity, and the control of gene regulation (Riechmann et al., 2000; Levine and Tjian, 2003). Phylogenetic studies have shown that the amplification and shuffling of protein domains have been determinant for the growth of certain transcription factor families (Laudet, 1997; Sharrocks, 2001; Ledent et al., 2002; Amoutzias et al., 2004) and different domains of a protein are often associated with different functions. (Baron et al., 1991; Sonnhammer and Kahn, 1994).

When dealing with transcription factor networks, several relevant questions arise, among them: How are these factors distributed and related through the network structure? How important are protein domains in shaping the network? An analysis of global patterns of network organization is required to answer these questions.

To this goal, here we explore the human transcription factor network (HTFN) obtained from the protein-protein interaction information from the TRANSFAC database (Wingender et al., 2001), using network analysis. We will show that this approximation allows us to obtain evolutionary footprints of the mechanisms shaping network architecture.

Data compilation from the TRANSFAC database provided 1 370 human entries. HTFN was built using a specific transcription factor database (TRANSFAC 8.2 professional database (Wingender et al., 2001)). We restricted our search to Homo sapiens using the database OS (organism) field. Information concerning physical interactions, derived from bibliographical sources, was extracted from the IN (interacting factor) database field. TRANSFAC contains, as entries, not only single transcription factors but also some entries for well-described transcription complexes. To avoid identifying a protein complex as a single protein, which could cause false or redundant interactions, we eliminated those complexes by selecting only entries with SQ field (protein sequence), which is only present in single transcription factors.

After filtering, a graph of $N = 230$ interacting human transcription factors was obtained (Figure 10.7a). The remaining transcription factors contained in the database did not form subgraphs and appeared isolated. The relatively small size of the connected graph compared with all the entries in the database might be due – at least in part – to the current degree of knowledge of this transcriptional regulatory network, with only sparse data for many of its components. Although a number of possible sources of bias are present, it is worth noting that the topological pattern of organization reported from different sources of protein-protein interactions seems consistent (Wagner, 2003).

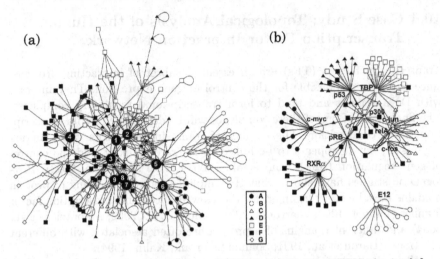

Fig. 10.7. Network representations of HTFN. (a) Entire HTFN obtained from TRANSFAC database. Numbers indicate hubs: 1, TATA binding protein (TBP); 2, p53; 3, p300; 4, retinoid X receptor α (RXRα); 5, retinoblastoma protein (pRB); 6, Nuclear factor NFκB p65 subunit (RelA); 7, c-jun; 8, c-myc; 9, c-fos. (b) k-scaffold of HTFN for a $k_c = 11$. Central box shows the symbols used for node representation accordingly the groups (A to G) defined by the topological overlap analysis (see Figure 10.9 and Table 10.2).

The topological analysis of HTFN is summarized in Table 10.1 revealing that HTFN is a sparse, small-world graph. The degree distribution (Figure 10.8a) and clustering (Figure 10.8b) exhibit a heterogeneous, skewed shape reminding us of a power-law behavior, thus indicating that most TFs are linked to only a few others whereas a handful of them (the hubs of HTFN) have many connections. The average betweenness centrality shows a well-defined power-law scaling (Figure 10.8c). Additionally, the network also displays well-defined correlations among proteins depending on their degree. As it occurs with other complex networks, we found that the HTFN is *disassortative*: Highly connected proteins attach to low connected ones (Newman, 2002a) and nodes with similar degree do not tend to be linked among them.

Figure 10.9a,b shows the obtained correlation profiles. They are similar to a protein interaction network of the yeast proteome (Maslov and Sneppen, 2002). The profile for HTFN (see Figure 10.9a), shows a dark region at the upper-left and bottom-right corners (positive Z value). This indicates that highly connected nodes associated with poorly-connected ones are much more abundant than predicted by the set of randomized networks. On the other hand, negative Z values in the domain representing relations between highly connected nodes (light regions in Figure 10.9a and b) indicate that they tend to be under-represented.

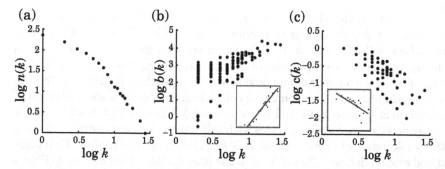

Fig. 10.8. Distributions for (a) degree, (b) betweenness centrality and (c) clustering. The power law fittings are shown in the insets. Linear regression coefficient: $r^2 = 0.96$ (a); betweenness centrality, $r^2 = 0.94$ (for b, inset); clustering coefficient $r^2 = 0.74$ (for c, inset).

Table 10.1. Topological parameters of some real networks: Human transcription factor network (HTFN); Erdös-Rényi (ER) null model network with N identical to that of the present study, proteome network from yeast (Jeong et al., 2001) and Internet (year 1999) (Vázquez et al., 2002; Newman, 2002b). For the ER model, we have used $\langle C \rangle = k/N$ and $\ell = log(N)/log\langle k \rangle$ (Newman, 2002a). Note that real networks are disassortative and exhibit a small-world pattern

	HTFN	ER model	Yeast proteome	Internet
N	230	230	1,870	10,100
L	851	851	4,488	38,380
$\langle k \rangle$	3.70	3.70	2.40	3.80
$\langle C \rangle$	0.17	0.015	0.07	0.24
ℓ	4.50	4.15	6.81	3.70
r	-0.18	-0.005	-0.15	-0.19

N: Total number of nodes; L: Total number of links; $\langle k \rangle$: Average degree; $\langle C \rangle$: Average clustering; ℓ: Average path length; r: Assortative mixing.

Scale-free networks exhibit a high degree of *error tolerance*, in other words, they are robust against random node deletions. They are, however, "vulnerable" to hub elimination (Albert et al., 2000). It seems that the effect of selective elimination has been attenuated in biomolecular networks by avoiding direct links between hubs (Maslov and Sneppen, 2002). This kind of pattern is associated with modularity: Groups of proteins can be identified as differentiated parts of the web, allowing functional diversity. Modularity can be properly detected and measured using the so-called *topological overlap matrix* (Ravasz et al., 2002). Figure 10.9c shows this overlap matrix for the HTFN. The array shows a nested, hierarchical structure, with small modules as dark boxes across the diagonal, which have a large overlap. However, there are some weak connections between modules, as shown by the tiny lines in the topological overlap matrix.

It is noteworthy that the presence of a high level of self-interactions is a prominent feature of this transcription factor web. In fact, 17.8% of proteins have self-interactions. Here, a self-interaction has to be understood as the interaction between proteins of the same type, i.e., homo-oligomerization, regardless of the number of monomers involved in such an interaction. To evaluate their importance, we compared correlation profiles with and without self-interactions (Figure 10.9a and b, respectively). Changes in the whole profile are evident, suggesting that nodes with self-interactions are distributed along the whole range of degree values. It is especially remarkable that the intense signal around degree values of $2-3$ in the profile with self-interactions (Figure 10.9a) is attenuated in the corresponding profile following their deletion (Figure 10.9b). Such striking differences can be explained by an over-abundance of proteins able to form either homo- or hetero-oligomers by means of connections with one or two more proteins (as it is the case of jun/jun, jun/fos, or myc/max/mad complexes (Hartl et al., 2003; Pelengaris and Khan, 2003)). These small clusters correspond to highly integrated modules depicted as small black regions in topological overlap matrix (Figure 10.9c). A simple explanation for these observations can be given based on biological constraints derived from the phylogeny of transcription factors, as will be discussed in the next section.

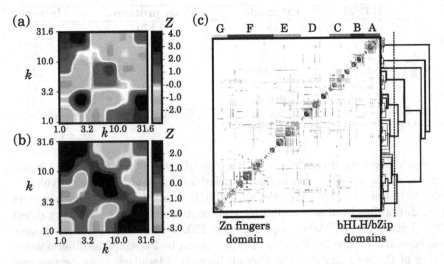

Fig. 10.9. Topological analysis of the HTFN. Here, the correlation profile is shown by using the Z-score taking into account self-interactions (a) and avoiding them (b). In (c), we display the topological overlap matrix and dendrogram. A to G indicate the topological groups defined by tracing a dashed line through the dendrogram. See Table 10.3 for biological and functional features of each group.

In order to clarify the relation between biological function and topology of HTFN, we identified in the network those factors that have the highest number of interactions (i.e., hubs). In a biological context, hubs can have important roles. Table 10.2 summarizes the most highly-connected factors in HTFN and their related diseases. They are also highlighted in the HTFN graph (Figure 10.7a). It should be stressed that the TATA binding protein (TBP) has the highest degree. TBP is considered a key factor for transcription initiation (Davidson, 2003). Other hubs, such as p53 (the second in degree) and retinoblastoma protein (pRB) are tumor suppressor proteins. Actually, most of these highly connected factors are related to cancer.

Table 10.2. Description and functionality of transcriptions factor hubs. Transcription factor (TF), degree (k), betweenness centrality (b).

TF	Description	Associated disease	k	$b \times 10^3$
TBP	Basal transcription machinery initiator	Spinocerebellar ataxia (Koide et al., 1999)	27	17.3
p53	Tumor suppressor protein	proliferative disease (Vousden and Prives, 2005)	23	18.5
p100	Coactivator. Histone acetyl-transferase	May play a role in epithelial cancer (Gayther et al., 2000)	18	20.2
RXRα	Retinoid X-α receptor	Hepatocellular carcinoma (Okuno et al., 2004)	18	8
pRB	Retinoblastoma suppressor protein. Tumor suppressor protein	Proliferative disease. Bladder cancer. Osteosarcoma (Liu et al., 2004)	15	27.1
RelA	NF-κB pathway	Hepatocyte apoptosis and fetal death (Joyce et al., 2001)	14	6.6
c-jun	AP-1 complex (activator). Proto-oncogen	Proliferative disease (Hartl et al., 2003)	14	4.1
c-myc	Activator. Proto-oncogen	Proliferative disease (Pelengaris and Khan, 2003)	13	10.5
c-fos	AP-1 complex (activator). Proto-oncogen	Proliferative disease (Sunters et al., 2004)	12	2

In order to reveal the mechanisms that shape the structure of HTFN, we studied its topological modularity in relation to the function and structure of transcription factors from available information. From a structural point of view, the over-abundance of self-interactions is associated with a majority group of 55% of basic helix-loop-helix (bHLH) and leucine zippers (bZip), a 17.5% of Zn fingers, and a 22.5% corresponding to a more heterogeneous group, the beta-scaffold factor with minor groove contact (according to the TRANSFAC classification) superclass, which includes Rel homology regions, MADS factors and others. Actually, such structures can be understood as protein domains, which can be found alone or combined to give rise to TFs. These domains are responsible for relevant properties, such as TF-DNA or

TF-TF binding. In this context, self-interactions can be explained by the presence of domains able to bind among them, as is the case of bHLH, bZip and the dimerizing regions in Zn finger nuclear receptor superfamily. They follow a general mechanism to interact with DNA based on protein dimerization (Branden and Tooze, 1999). According to this, the high clustering of the HTFN (see Figure 10.7) could be explained as a byproduct of the abundance of self-interacting domains.

We next wondered whether the HTFN modular architecture (Figure 10.9c) might include both functionality and structural similarity. In order to simplify the study of modularity, we traced an arbitrary line in the dendrogram of the topological overlap matrix, identifying seven putative protein groups (dashed line in Figure 10.9c). Nodes of each group have been identified by different symbols in the HTFN graph (see Figure 10.7a) where we can visualize the modules defined by the topological overlap algorithm. We must note that a consequence of the hierarchical component of HTFN is that not all factors in each group have the same level of relation.

Unlike a simple modular network, the combination of hierarchy and modularity cannot give purely homogeneous groups. Figure 10.7b shows the HTFN core graph, highlighting its modularity, the under-representation of connections between hubs and the overabundance of highly-connected nodes linked to poorly-connected ones (both observed in the correlation profile). The central role of the hubs in topological groups defined in Figure 10.7a should be stressed; these hubs are described in Table 10.2. An analysis of the topological modules of Figure 10.7 (labeled from A to G) shows that they include structural and/or functional features. Table 10.3 summarizes the main structural and functional features of these groups.

The methods concerning network analysis have been explained in the basic concepts section. We summarize here the ones concerning protein network acquisition.

10.5 Lessons Learned

The topological approach reveals that the HTFN exhibits a non-trivial organization where modularity and hierarchy play a fundamental role. Proteins with relevant biological roles occupy important positions in the network as hubs, thus indicating that node attributes (such as degree) can help elucidating the functional relevance of key elements in real networks.

Finally, we have shown that modularity is shaped by both structural and functional constrains and local rules of network growth based on tinkering largely determine global organization. These results allow defining possible evolutionary paths of network evolution.

Table 10.3. Structural and functional features of the groups obtained from topological overlap matrix.

Group & #TFs	Functional features	Functional features
A (22)	77% bHLH domains.	Muscle and neural tissue specific, sex determination. Includes E proteins family related to lymphocyte differentiation. (Quong et al., 2002; Morgenstern and Atchley, 1999). Includes E-box type A TF.
B (19)	47% bHLH-bZip domains.	c-myc related factors (59%). Includes E-box type B TF. Related to cell proliferation (Morgenstern and Atchley, 1999).
C (30)	36% rel homology region 40% bZip domains.	TF involved in NFκB pathway, AP1 complex and others
D (38)	24% fork head domains.	E2F/pRB pathway, histone deacetylases (HDAC) (Dimova and Dyson, 2005; Thiel et al., 2004). PRB and p53 isoforms...
E (45)	22% histone folding. Major part of specific interacting regions	Basal transcriptional machinery for promoters type I, II, III, PTF/SNAP complex and TBP related factors (Davidson, 2003; Lee and Young, 1998; Gangloff et al., 2001).
F (57)	42% Zn finger domains.	It contains the 90% of the members of nuclear receptor superfamily (they are Zn fingers also) of the HTFN.
G (19)	31% MAD domains.	SMAD family proteins and β-catenin and APC related factors.

10.6 List of Tools and Resources

We have presented some of the methods used in the analysis of complex networks. However, we can find other useful measures in the technical literature (Dorogovtsev and Mendes, 2003). In particular, alternatives have been suggested to quantify the network modularity. In this chapter, we have presented a method based on the clustering property, but modularity can also be measured by means of betweenness (Palla et al., 2005; Radicchi et al., 2004).

The k-scaffold graph analysis is an alternative to the so-called k-core (Seidman, 1983), which allows to retain only those nodes with at least n edges; edges connecting these nodes are not displayed.

For graph visualization, the most popular visualization software is *Pajek* (http://vlado.fmf.uni-lj.si/pub/networks/pajek/), which is free for Windows operating systems. *Pajek* provides a graphic interface and a set of algorithms for graph analysis. In our study, we used the *Graphviz* package (http://www.graphviz.org/). Recently, path visualization tools are being incorporated in biological databases offering molecule information in the context of the interactions with other molecules. This is the case for commercial databases such as *Protein lounge* (www.proteinlounge.com/), Transfac (http://www.biobase.de/). In this context, we can also mention *Cytoscape* (http://www.cytoscape.org/), a bioinformatics software platform for visualizing molecular interaction networks and integrating them with gene expression profiles and other data. This system allows to integrate interacting databases and extract gene functional annotation databases from gene ontology.

10.7 Conclusions

The study of large-scale cellular networks based on graph theory has revealed that proteins and metabolites are connected in a non-trivial way. The network approach reveals that, although biological functions originate from the interaction among specific elements, global organization exhibits universal topological properties. Real networks are often sparse, scale-free, hierarchical and exhibit a small-world pattern. Experimental systemic approaches, such as high-throughput methods could benefit from this type of integrative views in order to gain a better understanding of massive biological data. The case study presented here is an example of such an application. The HTFN analysis shows that this network shares topological properties with other real graphs. We have shown that the highly connected nodes are related to essential functions and that topological features, such as modularity, capture information concerning the functionality and phylogeny of its components.

10.8 Mathematical Details

Assortative mixing (r). Given the degree distribution p_k of the studied graph G_Ω, the expected probability of two arbitrary nodes of degrees $k_0 = m$ and $k_1 = n$ to be linked in the *uncorrelated network* counterpart (with the same p_k as G_Ω) is the product of their degree abundance, that is, $q_m \times q_n$. The probability that a link connects two nodes with degrees m and n in G_Ω is denoted as p_{mn}. It is calculated dividing the number of times that two nodes with m and n degrees are connected in G_Ω by the total number of possibilities $(N \times N)$. Then, we define a global correlation function weighting the degree for each correlation $q_m q_n$ and p_{mn}, as $C(m,n) = \langle mn \rangle - \langle m \rangle \langle n \rangle = \sum_{m,n} p_{nm} - \sum_{mn} mn q_m q_n$.

To compare different networks we show the expression for the normalized correlation function (Fernández and Solé, 2005). Normalization is achieved by using the variance of the degree distribution σ_q^2 of G_Ω, $\sum_k k^2 q_k - [\sum_k k q_k]^2$, Hence the normalized correlation function is expressed as

$$r = \frac{1}{\sigma_q^2} \sum_{mn} mn(p_{mn} - q_m q_n) \tag{10.4}$$

To calculate r, this expression is formulated as follows: (Newman, 2002b)

$$r = \frac{L^{-1} \sum_i j_i k_i - [L^{-1} \sum_i \frac{1}{2}(j_i + k_i)]^2}{L^{-1} \sum \frac{1}{2}(j_i^2 + k_i^2) - [L^{-1} \sum \frac{1}{2}(j_i + k_i)]^2}, \qquad (10.5)$$

where j_i, k_i are the degrees of the vertices at the ends of the ith edge, with $i = 1, ... L$.

Acknowledgments

This work was supported by grants FIS2004-05422 and the National Institute of Health (CA 113004).

References

Albert, R. (2005). Scale-free networks in cell biology. *J. Cell Sci.*, 118(Pt 21):4947–57.

Albert, R. and Barabasi, A. L. (2002). Statistical mechanics of complex networks. *Rev. Modern Phys.*, 74:47–97.

Albert, R., Jeong, H., and Barabasi, A. L. (1999). Diameter of the world-wide web. *Nature*, 401:130–131.

Albert, R., Jeong, H., and Barabasi, A. L. (2000). Error and attack tolerance of complex networks. *Nature*, 406(6794):378–82.

Amaral, L.A.N., Scala, A., Barthél'emy, M., and H.E., Standley (2000). Classes of small-world networks. *Proc. Natl. Acad. Sci. USA*, 97:11149–11152.

Amoutzias, G. D., Robertson, D. L., Oliver, S. G., and Bornberg-Bauer, E. (2004). Convergent networks by single-gene duplications in higher eukaryotes. *EMBO Rep.*, 5(3):274–9.

Barabasi, A.L. and Oltvai, Z.N. (2004). Network biology: Understanding the cell's functional organization. *Nat. Rev. Genet.*, 5(2):101–13.

Baron, M., Norman, D.G., and Campbell, I.D. (1991). Protein modules. *Trends Biochem. Sci.*, 16(1):13–17.

Branden, C. and Tooze, J. (1999). *Introduction to Protein Structure*. Garland Publishing, Inc., New York.

Davidson, I. (2003). The genetics of tbp and tbp-related factors. *Trends Biochem. Sci.*, 28(7):391–8.

Dimova, D. K. and Dyson, N. J. (2005). The E2F transcriptional network: Old acquaintances with new faces. *Oncogene*, 24(17):2810–26.

Dorogovtsev, S. N. and Mendes, J. F. F. (2003). *Evolution of networks: from biological nets to the internet and WWW*. Oxford University Press., Oxford.

Erdös, P. and Rényi, A. (1960). On the evolution of random graphs. *Publ. Math. Inst. Hung. Acad. Sci.*, 5:17–60.

Fernández, P. and Solé, R. V. (2005). Graphs as models of large-scale biochemical organization. In Bonchev, D. and Rouvray, D. H., editors, *Complexity in chemistry, biology and ecology*. Springer, New York.

Ferrer, R. and Solé, R.V. (2001). The small world of human language. *Proc. Roy. Soc. Lond. B*, 268:2261–2265.

Gangloff, Y. G., Romier, C., Thuault, S., Werten, S., and Davidson, I. (2001). The histone fold is a key structural motif of transcription factor tfiid. *Trends Biochem. Sci.*, 26(4):250–7.

Gavin, A. C., Bosche, M., and Krause, R. et al. (2002). Functional organization of the yeast proteome by systematic analysis of protein complexes. *Nature*, 415(6868):141–7.

Gayther, S. A., Batley, S. J., and Linger, L. et al. (2000). Mutations truncating the EP300 acetylase in human cancers. *Nat. Genet.*, 24(3):300–3.

Giot, L., Bader, J. S., and Brouwer, C. et al. (2003). A protein interaction map of drosophila melanogaster. *Science*, 302(5651):1727–36.

Guelzim, N., Bottani, S., Bourgine, P., and Kepes, F. (2002). Topological and causal structure of the yeast transcriptional regulatory network. *Nat. Genet.*, 31(1):60–3.

Han, J. D., Dupuy, D., Bertin, N., Cusick, M. E., and Vidal, M. (2005). Effect of sampling on topology predictions of protein-protein interaction networks. *Nat. Biotechnol.*, 23(7):839–44.

Hartl, M., Bader, A.G., and Bister, K. (2003). Molecular targets of the oncogenic transcription factor jun. *Curr. Cancer Drug Targets*, 3(1):41–55.

Ito, T., Chiba, T., Ozawa, R., Yoshida, M., Hattori, M., and Sakaki, Y. (2001). A comprehensive two-hybrid analysis to explore the yeast protein interactome. *Proc. Natl. Acad. Sci. USA*, 98(8):4569–74.

Jeong, H., Mason, S.P., Barabasi, A.L., and Oltvai, Z.N. (2001). Lethality and centrality in protein networks. *Nature*, 411(6833):41–2.

Jeong, H., Tombor, B., Albert, R., Oltvai, Z. N., and Barabasi, A. L. (2000). The large-scale organization of metabolic networks. *Nature*, 407(6804):651–4.

Joyce, D., Albanese, C., Steer, J., Fu, M., Bouzahzah, B., and Pestell, R.G. (2001). Nf-κb and cell-cycle regulation: the cyclin connection. *Cytokine Growth Factor Rev*, 12(1):73–90.

Koide, R., Kobayashi, S., Shimohata, T., Ikeuchi, T., Maruyama, M., Saito, M., Yamada, M., Takahashi, H., and Tsuji, S. (1999). A neurological disease caused by an expanded cag trinucleotide repeat in the tata-binding protein gene: a new polyglutamine disease? *Hum. Mol. Genet.*, 8(11):2047–53.

Laudet, V. (1997). Evolution of the nuclear receptor superfamily: early diversification from an ancestral orphan receptor. *J. Mol. Endocrinol.*, 19(3):207–26.

Ledent, V., Paquet, O., and Vervoort, M. (2002). Phylogenetic analysis of the human basic helix-loop-helix proteins. *Genome Biol.*, 3(6):RE-SEARCH0030.

Lee, T.I., Rinaldi, N.J., and Robert, F. et al. (2002). Transcriptional regulatory networks in saccharomyces cerevisiae. *Science*, 298(5594):799–804.

Lee, T.I. and Young, R.A. (1998). Regulation of gene expression by tbp-associated proteins. *Genes Dev.*, 12(10):1398–408.

Levine, M. and Tjian, R. (2003). Transcription regulation and animal diversity. *Nature*, 424(6945):147–51.

Li, S., Armstrong, C. M., Bertin, N., Ge, H., Milstein, S., Boxem, M., Vidalain, P. O., Han, J. D., Chesneau, A., Hao, T., Goldberg, D. S., Li, N., Martinez, M., Rual, J. F., Lamesch, P., Xu, L., Tewari, M., Wong, S. L., Zhang, L. V., Berriz, G. F., Jacotot, L., Vaglio, P., Reboul, J., Hirozane-Kishikawa, T., Li, Q., Gabel, H. W., Elewa, A., Baumgartner, B., Rose, D. J., Yu, H., Bosak, S., Sequerra, R., Fraser, A., Mango, S. E., Saxton, W. M., Strome, S., Van Den Heuvel, S., Piano, F., Vandenhaute, J., Sardet, C., Gerstein, M., Doucette-Stamm, L., Gunsalus, K. C., Harper, J. W., Cusick, M. E., Roth, F. P., Hill, D. E., and Vidal, M. (2004). A map of the interactome network of the metazoan c. elegans. *Science*, 303(5657):540–3.

Liu, H., Dibling, B., Spike, B., Dirlam, A., and Macleod, K. (2004). New roles for the rb tumor suppressor protein. *Curr. Opin. Genet. Dev.*, 14(1):55–64.

Ma'ayan, A., Jenkins, S.L., Neves, S., Hasseldine, A., Grace, E., Dubin-Thaler, B., Eungdamrong, N.J., Weng, G., Ram, P.T., Rice, J.J., Kershenbaum, A., Stolovitzky, G.A., Blitzer, R.D., and Iyengar, R. (2005). Formation of regulatory patterns during signal propagation in a mammalian cellular network. *Science*, 309(5737):1078–83.

Maslov, S. and Sneppen, K. (2002). Specificity and stability in topology of protein networks. *Science*, 296(5569):910–3.

McCraith, S., Holtzman, T., Moss, B., and Fields, S. (2000). Genome-wide analysis of vaccinia virus protein-protein interactions. *Proc. Natl. Acad. Sci. USA*, 97(9):4879–84.

Milgram, S. (1967). The small-world problem. *Psychol. Today*, 2:60–67.

Milo, R., Shen-Orr, S., Itzkovitz, S., Kashtan, N., Chklovskii, D., and Alon, U. (2002). Network motifs: Simple building blocks of complex networks. *Science*, 298(5594):824–7.

Morgenstern, B. and Atchley, W. R. (1999). Evolution of BHLH transcription factors: modular evolution by domain shuffling? *Mol. Biol. Evol.*, 16(12):1654–63.

Newman, M. (2002a). Random graphs as models of networks. In Bornholdt, S. and Schuster, H.G., editors, *Handbook of Graphs and Networks*. Wiley-VHC, Weinheim.

Newman, M. E. (2002b). Assortative mixing in networks. *Phys. Rev. Lett.*, 89(20):208701.

Okuno, M., Kojima, S., Matsushima-Nishiwaki, R., Tsurumi, H., Muto, Y., Friedman, S. L., and Moriwaki, H. (2004). Retinoids in cancer chemoprevention. *Curr. Cancer Drug Targets*, 4(3):285–98.

Ouzounis, C. A. and Karp, P. D. (2000). Global properties of the metabolic map of escherichia coli. *Genome Res.*, 10(4):568–76.

Palla, G., Derenyi, I., Farkas, I., and Vicsek, T. (2005). Uncovering the overlapping community structure of complex networks in nature and society. *Nature*, 435(7043):814–8.

Pastor-Satorras, R., Smith, E., and Solé, R.V. (2003). Evolving protein interaction networks through gene duplication. *J. Theor. Biol.*, 222(2):199–210.

Pelengaris, S. and Khan, M. (2003). The many faces of c-myc. *Arch. Biochem. Biophys.*, 416(2):129–36.

Ptacek, J., Devgan, G., Michaud, G., Zhu, H., Zhu, X., Fasolo, J., Guo, H., Jona, G., Breitkreutz, A., Sopko, R., McCartney, R.R., Schmidt, M.C., Rachidi, N., Lee, S.J., Mah, A.S., Meng, L., Stark, M.J., Stern, D.F., De Virgilio, C., Tyers, M., Andrews, B., Gerstein, M., Schweitzer, B., Predki, P.F., and Snyder, M. (2005). Global analysis of protein phosphorylation in yeast. *Nature*, 438(7068):679–84.

Quong, M. W., Romanow, W. J., and Murre, C. (2002). E protein function in lymphocyte development. *Annu. Rev. Immunol.*, 20:301–22.

Radicchi, F., Castellano, C., Cecconi, F., Loreto, V., and Parisi, D. (2004). Defining and identifying communities in networks. *Proc. Natl. Acad. Sci. USA*, 101(9):2658–63.

Rain, J.C., Selig, L., De Reuse, H., Battaglia, V., Reverdy, C., Simon, S., Lenzen, G., Petel, F., Wojcik, J., Schachter, V., Chemama, Y., Labigne, A., and Legrain, P. (2001). The protein-protein interaction map of *Helicobacter pylori*. *Nature*, 409(6817):211–5.

Ravasz, E., Somera, A.L., Mongru, D.A., Oltvai, Z.N., and Barabasi, A.L. (2002). Hierarchical organization of modularity in metabolic networks. *Science*, 297(5586):1551–5.

Riechmann, J.L., Heard, J., Martin, G., Reuber, L., Jiang, C., Keddie, J., Adam, L., Pineda, O., Ratcliffe, O.J., Samaha, R.R., Creelman, R., Pilgrim, M., Broun, P., Zhang, J.Z., Ghandehari, D., Sherman, B.K., and Yu, G. (2000). *Arabidopsis* transcription factors: Genome-wide comparative analysis among eukaryotes. *Science*, 290(5499):2105–10.

Rodriguez-Caso, C., Medina, M.A., and Solé, R.V. (2005). Topology, tinkering and evolution of the human transcription factor network. *FEBS J.*, 272(24):6423–34.

Rual, J.F., Venkatesan, K., Hao, T., Hirozane-Kishikawa, T., Dricot, A., Li, N., Berriz, G.F., Gibbons, F.D., Dreze, M., Ayivi-Guedehoussou, N., Klitgord, N., Simon, C., Boxem, M., Milstein, S., Rosenberg, J., Goldberg, D.S., Zhang, L. V., Wong, S.L., Franklin, G., Li, S., Albala, J.S., Lim, J., Fraughton, C., Llamosas, E., Cevik, S., Bex, C., Lamesch, P., Sikorski, R.S., Vandenhaute, J., Zoghbi, H.Y., Smolyar, A., Bosak, S., Sequerra, R., Doucette-Stamm, L., Cusick, M.E., Hill, D.E., Roth, F.P., and Vidal, M. (2005). Towards a proteome-scale map of the human protein-protein interaction network. *Nature*, 437(7062):1173–8.

Salgado, H., Gama-Castro, S., Martinez-Antonio, A., Diaz-Peredo, E., Sanchez-Solano, F., Peralta-Gil, M., Garcia-Alonso, D., Jimenez-Jacinto, V., Santos-Zavaleta, A., Bonavides-Martinez, C., and Collado-Vides, J.

(2004). RegulonDB (version 4.0): transcriptional regulation, operon organization and growth conditions in *Escherichia coli* K-12. *Nucleic Acids Res.*, 32(Database issue):D303–6.

Seidman, S. B. (1983). Network structure and minimum degree. *Social Networks*, 5:269–287.

Sharrocks, A.D. (2001). The ETS-domain transcription factor family. *Nat. Rev. Mol. Cell Biol.*, 2(11):827–37.

Shen-Orr, S.S., Milo, R., Mangan, S., and Alon, U. (2002). Network motifs in the transcriptional regulation network of *Escherichia coli*. *Nat. Genet.*, 31(1):64–8.

Solé, R.V. and Montoya, J.M. (2001). Complexity and fragility in ecological networks. *Proc. Roy. Soc. Lond. B Biol. Sci.*, 268(1480):2039–45.

Solé, S.V. and Valverde, S. (2006). Are networks motifs the spandrels of cellular complexity? *TREE*, 21(8):419–422.

Sonnhammer, E.L. and Kahn, D. (1994). Modular arrangement of proteins as inferred from analysis of homology. *Protein Sci.*, 3(3):482–92.

Stelzl, U., Worm, U., Lalowski, M., Haenig, C., Brembeck, F.H., Goehler, H., Stroedicke, M., Zenkner, M., Schoenherr, A., Koeppen, S., Timm, J., Mintzlaff, S., Abraham, C., Bock, N., Kietzmann, S., Goedde, A., Toksoz, E., Droege, A., Krobitsch, S., Korn, B., Birchmeier, W., Lehrach, H., and Wanker, E.E. (2005). A human protein-protein interaction network: a resource for annotating the proteome. *Cell*, 122(6):957–68.

Sunters, A., Thomas, D.P., Yeudall, W.A., and Grigoriadis, A.E. (2004). Accelerated cell cycle progression in osteoblasts overexpressing the c-fos proto-oncogene: induction of cyclin a and enhanced cdk2 activity. *J. Biol. Chem.*, 279(11):9882–91.

Tanaka, J. (2005). Scale-rich metabolic networks. *Phys. Rev. Lett.*, 94:168101.

Thiel, G., Lietz, M., and Hohl, M. (2004). How mammalian transcriptional repressors work. *Eur. J. Biochem.*, 271(14):2855–62.

Uetz, P., Giot, L., Cagney, G., Mansfield, T.A., Judson, R.S., Knight, J.R., Lockshon, D., Narayan, V., Srinivasan, M., Pochart, P., Qureshi-Emili, A., Li, Y., Godwin, B., Conover, D., Kalbfleisch, T., Vijayadamodar, G., Yang, M., Johnston, M., Fields, S., and Rothberg, J.M. (2000). A comprehensive analysis of protein-protein interactions in *Saccharomyces cerevisiae*. *Nature*, 403(6770):623–7.

Uetz, P. and Hughes, R. E. (2000). Systematic and large-scale two-hybrid screens. *Curr. Opin. Microbiol.*, 3(3):303–8.

Vázquez, A., Pastor-Satorras, R., and Vespignani, A. (2002). Large-scale topological and dynamical properties of the internet. *Phys. Rev. E Stat. Nonlin. Soft Matter Phys.*, 65(6 Pt 2):066130.

Vousden, K. H. and Prives, C. (2005). P53 and prognosis: New insights and further complexity. *Cell*, 120(1):7–10.

Wagner, A. (2003). How the global structure of protein interaction networks evolves. *Proc. Biol. Sci.*, 270(1514):457–66.

Wagner, A. and Fell, D. A. (2001). The small world inside large metabolic networks. *Proc. Biol. Sci.*, 268(1478):1803–10.

Walhout, A.J., Boulton, S.J., and Vidal, M. (2000). Yeast two-hybrid systems and protein interaction mapping projects for yeast and worm. *Yeast*, 17(2):88–94.

Watts, D. J. and Strogatz, S. H. (1998). Collective dynamics of 'small-world' networks. *Nature*, 393(6684):440–2.

Wingender, E., Chen, X., Fricke, E., Geffers, R., Hehl, R., Liebich, I., Krull, M., Matys, V., Michael, H., Ohnhauser, R., Pruss, M., Schacherer, F., Thiele, S., and Urbach, S. (2001). The TRANSFAC system on gene expression regulation. *Nucleic Acids Res.*, 29(1):281–3.

Wyrick, J.J. and Young, R.A. (2002). Deciphering gene expression regulatory networks. *Curr. Opin. Genet. Dev.*, 12(2):130–6.

Identifying Important Explanatory Variables for Time-Varying Outcomes

Oliver Bembom, Maya L. Petersen, and Mark J. van der Laan

Division of Biostatistics, University of California, Berkeley, USA.
bembom@berkeley.edu, mayaliv@berkeley.edu, laan@stat.berkeley.edu

11.1 Introduction

Many applications in modern biology measure a large number of genomic or proteomic covariates and are interested in assessing the impact of each of these covariates on a particular outcome of interest. In a study which follows a cohort of HIV-positive patients over time, for example, a researcher may genotype the virus infecting each patient to ascertain the presence or absence of a large number of mutations, in the hope of identifying mutations that affect how a patient's plasma HIV RNA level (viral load) responds to a new drug regimen. Along with an estimate of the impact of each mutation on the time course of viral load, the researcher would generally like to have a measure of the statistical significance of these estimates in order to identify those mutations that are most likely to be genuinely related to the outcome. Such information could then be used to inform the decision of which drugs should be included in the regimen of a patient with a particular pattern of mutations.

To tackle this problem, we first need to define precisely what we mean by "the impact of a mutation on the time course of viral load". For this purpose, let us denote the collection of candidate mutations by $A = (A_1, \ldots, A_p)$, with $A_j = 1$ if a specific amino acid substitution is present at the given position and $A_j = 0$ otherwise. Let $Y(t)$ denote a patient's viral load measured at time t. Suppose we also measure a number of clinical covariates $C = (C_1, \ldots, C_q)$ at baseline that tend to be associated with the occurrence of particular mutations and that independently affect a patient's virologic response.

The simplest way of assessing the impact of a particular mutation A_j on $Y(t)$ would now be to compare the virologic response among patients with $A_j = 1$ to that among patients with $A_j = 0$. If we find that patients in the first group respond much more poorly to a particular drug regimen, a clinician might be inclined not to give this regimen to a new patient entering his office who has this mutation. Patients in the first group are, however, also quite likely to differ from those in the second group in terms of the remaining mutations and the clinical covariates C. The mutation A_j may, for example, be very common among patients who have previously failed several similar drug regimens, making them far more likely to also fail the current one, but very rare among other patients. If the clinician's new patient comes from a population that differs from our original study population in that the mutation is not associated with having previously failed similar drug regimens, we might be wrong to conclude that the regimen under consideration would be a poor choice in this situation. Since the impact of A_j on $Y(t)$ is *confounded* by the clinical baseline covariates C, our results do not generalize to a new population in which A_j and C are related to each other in a different way.

We might thus be interested in estimating the impact of A_j on $Y(t)$ that is not due to associations of A_j with any of the baseline covariates C. Specifically, we might ask: What difference in virologic response would we observe if we could somehow give every patient in our study population the mutation A_j, *holding their clinical covariates C fixed at their current values*, as opposed to the scenario in which we give none of the patients this mutation, holding again C fixed? Any observed difference could then not be due to differences of the two populations with regard to C and would thus be more likely to generalize to a new population in which A_j and C may be related to each other differently.

To appreciate the difference between the estimates obtained in this way and those described earlier, consider the ideal experiment that would correspond to these earlier estimates. If we simply compare patients with $A_j = 1$ to those with $A_j = 0$, we would be asking: What difference in virologic response would we observe if we gave every patient in our study population the mutation A_j, *allowing their clinical covariates to take on values that are typical for patients with $A_j = 1$*, as opposed to the scenario in which we give none of the patients this mutation, again allowing C to take on typical values? If we now encounter a patient from a new population, the typical values of C for patients with $A_j = 1$ that we observed in our original study population may not correspond to typical values of C for such patients in this new population.

In the hypothetical experiment in which we control for C by holding it fixed at its observed values, any other covariates that are not included in C are implicitly allowed to take on values typical for the value of A_j we are considering. In particular, some of the remaining mutations may be strongly correlated with A_j so that they would be likely to change their values if we assigned every patient $A_j = 1$ or $A_j = 0$. If these other mutations are now themselves independently related to $Y(t)$, our estimates for the impact of A_j on $Y(t)$ may not translate well to a new population in which the mutations tend to occur in somewhat different patterns. Only if we adjust for all confounders of the relationship between A_j and $Y(t)$, i.e., all covariates that are associated with A_j and that are functionally related to $Y(t)$, can we be sure that our estimates will be applicable to a new population of patients. If we do so, we are in fact estimating the *causal* impact of A_j on $Y(t)$, rather than a mere association between A_j and $Y(t)$.

We note, however, that estimates of the impact of A_j on $Y(t)$ may be interesting and meaningful even if we are not in the ideal situation of being able to adjust for all relevant confounders. We can still identify mutations that are strongly associated with $Y(t)$ and that may thus allow us to predict a new patient's virologic response to a particular drug regimen, assuming that this patient comes from a population in which the unmeasured confounders are associated with A_j in a manner that is not too dissimilar from that observed in our study population. Depending on the nature of the unmeasured confounders, this assumption may not be at all unreasonable.

In this chapter, we describe an approach that allows us to estimate such measures of variable importance for any set of baseline covariates we may wish to adjust for. Mathematically speaking, these methods allow us to estimate the parameter

$$\Psi_j(t) = E\Big[E[Y(t) \mid A_j = 1, W_j] - E[Y(t) \mid A_j = 0, W_j]\Big] \qquad (11.1)$$

for each j and each t, where $W_j = (W_j^1, \ldots, W_j^m)$ is the desired set of adjustment variables. The estimates we obtain rely on a minimum number of assumptions and in some cases are as precise as possible. Furthermore, the approach provides an honest measure of the statistical significance of each estimate. For a more rigorous treatment, the interested reader is referred to the article by (van der Laan, 2006b).

11.2 Basic Concepts

In this section, we describe how the variable importance parameter $\Psi_j(t)$ is estimated in practice. The central step consists of transforming the recorded data for each observation into a quantity whose expectation equals $\Psi_j(t)$. Conceptually, these quantities can be thought of as giving a measure of the impact of A_j on $Y(t)$ as derived from a single observation. We can then estimate the entire function $\Psi_j(t)$ relating variable importance to time by fitting a statistical model for how the expectation of the transformed quantities depends on time, i.e., by *regressing* them on time.

We will describe three different transformations of the observed data that are suitable for our purposes. These transformations themselves involve parameters that are generally not known by the researcher and hence must be estimated from the observed data. Since these parameters are not of interest in themselves, we refer to them as *nuisance parameters*. The first of these nuisance parameters is the so-called treatment mechanism. The variable A_j whose effect on $Y(t)$ we would like to estimate is often referred to as the treatment variable. The treatment mechanism g_j now gives the probability of observing a given treatment $A_j = a$ for a subject with a particular covariate profile W_j:

$$g_j(a, W_j) \equiv P(A_j = a \mid W_j) \qquad (11.2)$$

The second nuisance parameter consists of a regression of $Y(t)$ on A_j and W_j:

$$Q_j(a, t, W_j) \equiv E[Y(t) \mid A_j = a, W_j] \qquad (11.3)$$

To estimate g_j and Q_j we ideally do not want to rely on the assumption of a particular functional form. For example, we would like to avoid an assumption such as that the expectation of $Y(t)$ given A_j and W_j can be written as

$$E[Y(t) \mid A_j, W_j] = \beta_{j0} + \beta_{j1} A_j + \beta_{j2} t + \beta_{j3} W_j^1 + \ldots + \beta_{j(m+2)} W_j^m \quad (11.4)$$

for some coefficient vector $\beta_j = (\beta_{j0}, \ldots, \beta_{j(m+2)})$. On the basis of biological knowledge alone it is very difficult to arrive at an appropriate functional form, and poorly specified functional forms can lead to severely biased estimates of variable importance. Thus, the functional form of the nuisance parameter models should be chosen based on the information that is contained in the data set, i.e., data-adaptively. One popular approach to this model selection problem is the D/S/A algorithm (Sinisi and van der Laan, 2004) that relies on deletion, substitution, and addition moves to search through a large space of possible functional forms. Another well-known model selection technique that searches through a much smaller space of candidate functional forms and thus requires somewhat less computing time has been introduced by Kooperberg et al. (1997).

In modeling Q, we have to bear in mind that repeated viral load measurements on the same patient will be correlated. The usual generalized linear models were formulated for outcomes that are independent of each other (McCullagh and Nelder, 1989). In the presence of correlated outcomes, these models provide estimates that, while still reliable, are no longer as precise as possible. Furthermore, the p-values they provide for testing whether or not certain regression coefficients are equal to zero cannot be trusted since they tend to be too small.

Generalized estimating equations address both of these issues and have thus been a popular tool for modeling correlated outcomes (Liang and Zeger, 1986; Zeger and Liang, 1986). Unlike generalized estimating equations, the D/S/A algorithm for data-adaptive model selection does not explicitly take into account the correlation between outcomes $Y(t)$ measured on the same subject. Although the model fits obtained by this algorithm are not as precise as possible, the estimates are still reliable and useful for the purpose of model selection. Furthermore, the regression of $Y(t)$ on A_j and W_j is only a nuisance parameter, needed to estimate the variable importance of A_j, but is not of primary interest in itself. Thus, we are not interested in testing whether some coefficients in the selected model might be equal to zero or not. The only adjustment that is necessary when using the D/S/A algorithm in this situation is to supply an ID variable that can be used to identify the independent experimental units. This allows the algorithm to carry out an honest cross-validation procedure by assigning measurements from the same subject to the same validation sample. The approach by Kooperberg et al. (1997) can be used for modeling the treatment mechanism, but cannot be extended in a straightforward way for model selection in the context of correlated outcomes.

Given estimates $g_{j,n}$ and $Q_{j,n}$ of the nuisance parameters g_j and Q_j, we can now generate three different types of transformations of the observed data. Let $T_{i,k}$ denote the time at which the kth measurement of the outcome Y was obtained for subject i. For each outcome measurement $Y_{i,k}(T_{i,k})$ for subject i, we will obtain one transformed observation $D_{i,k}^j$. The regression-based transformation only makes use of the nuisance parameter Q_j and is given by

$$D_{i,k}^j(T_{i,k}) = Q_{j,n}(1, T_{i,k}, W_{j,i}) - Q_{j,n}(0, T_{i,k}, W_{j,i}) \qquad (11.5)$$

The inverse-probability-of-treatment-weighted (IPTW) transformation only makes use of the nuisance parameter g_j and is given by

$$D_{i,k}^j(T_{i,k}) = \left\{ \frac{I(A_{j,i} = 1)}{g_{j,n}(1, W_{j,i})} - \frac{I(A_{j,i} = 0)}{g_{j,n}(0, W_{j,i})} \right\} Y_{i,k}(T_{i,k}) \qquad (11.6)$$

where $I(\cdot)$ is the indicator function that equals one if the condition in parentheses is true and zero otherwise. The double robust transformation, finally, makes use of both nuisance parameters and is given by

$$\begin{aligned} D_{i,k}^j(T_{i,k}) = {} & Q_{j,n}(1, T_{i,k}, W_{j,i}) - Q_{j,n}(0, T_{i,k}, W_{j,i}) + \\ & \left\{ \frac{I(A_{j,i} = 1)}{g_{j,n}(1, W_{j,i})} \Big(Y_{i,k}(T_{i,k}) - Q_{j,n}(1, T_{i,k}, W_{j,i}) \Big) - \right. \\ & \left. \frac{I(A_{j,i} = 0)}{g_{j,n}(0, W_{j,i})} \Big(Y_{i,k}(T_{i,k}) - Q_{j,n}(0, T_{i,k}, W_{j,i}) \Big) \right\} \end{aligned} \qquad (11.7)$$

Since both the regression-based and IPTW transformation of the data only rely on one of the two estimated nuisance parameters, variable importance estimates based on these transformations will only be reliable if the relevant nuisance parameter is estimated well. The double robust transformation, however, relies on both nuisance parameters and has the remarkable property that it yields correct estimates of variable importance if either one of these two nuisance parameters is estimated well.

We now obtain three different estimates of the variable importance parameter $\Psi_j(t)$ by regressing each of these transformed observations on time t. As above, the use of data-adaptive model selection approaches such as the D/S/A algorithm for fitting these regressions can help to minimize the reliance on assumptions about how variable importance varies as a function of time. The same approach can be used to obtain estimates of the variable importance of a covariate A_j conditional on some subset V of the adjustment covariates W_j. For example, a researcher may be interested in the impact of a given mutation on the time course of viral load conditional on the viral load at baseline. Mathematically speaking, this corresponds to the parameter

$$\Psi_j(t, v) = E\left[E[Y(t) \mid A_j = 1, W_j] - E[Y(t) \mid A_j = 0, W_j] \Big| V = v\right] \quad (11.8)$$

Such parameters are straightforward to estimate once we have created the transformed observations described above by simply regressing them on both t and V rather than on t alone.

Plots of the estimated variable importance $\Psi_{j,n}(t)$ as a function of time can now be used to explore how the impact of A_j on $Y(t)$ changes over time. Such plots can furthermore be used as inputs for a clustering algorithm to identify treatment variables A_j whose impact on $Y(t)$ develops according to a similar dynamic over time. Alternatively, we may be interested in testing the hypotheses that the importance of treatment variable A_j is zero at some set of time points $t_1, \ldots, t_d$, with the goal of identifying treatment variables A_j^* and time points t^* for which we have strong evidence against such hypotheses. For this purpose, we use the following bootstrap approach to first obtain separate p-values $p_{j,1}, \ldots, p_{j,d}$ for the respective hypotheses that $\Psi_j(t_1), \ldots, \Psi_j(t_d)$ equal zero. We draw a large number of samples of size n with replacement from the pool of n subjects in our data set to obtain bootstrap data sets that contain all outcome measurements $Y(t)$ for the selected subjects. For each of these bootstrap samples we now repeat the entire estimation process as outlined above to arrive at bootstrap estimates $\Psi_{j,n}^{\#}(t_1), \ldots, \Psi_{j,n}^{\#}(t_d)$ of the desired variable importance measures. If we have used a computationally involved model selection technique like the D/S/A algorithm to regress A_j on W_j, $Y(t)$ on A_j and W_j, or $D(t)$ on t, we may avoid the model selection step as part of this bootstrap process and simply refit the regressions according to the selected functional form. This approach saves a significant amount of time and generally leads to p-values that are only slightly optimistic.

We can now take the variance of these bootstrap estimates as an estimate of the variance of $\Psi_{j,n}(t_k)$ under the null hypothesis that $\Psi_j(t_k) = 0$ and form t-statistics by dividing $\Psi_{j,n}(t_k)$ by the square root of this estimated variance. Under the null hypotheses, these test statistics will be closely approximated by a standard normal distribution once we have a reasonable sample size. The desired p-value can thus be obtained as the probability that the absolute value of a standard normal variate exceeds the absolute value of the observed test statistic. Each of these p-values gives an estimate of the proportion of times we would reject a true null hypothesis if the experiment and corresponding hypothesis test were to be performed over and over again. The p-values are formulated for individual hypothesis tests and thus do not take into account that we are testing several hypotheses simultaneously.

A large number of methods exist for obtaining p-values that are interpretable in this context of multiple testing (Westfall and Young, 1993; Lehmann and Romano, 2005; Dudoit and van der Laan, 2006). Among the most straightforward methods are those that simply transform the raw p-values obtained from the individual hypothesis tests into a set of adjusted p-values. The well-known Bonferroni adjustment, for example, simply multiplies each p-value by the number of comparisons that are made (Bland and

Altman, 1995). The adjusted p-values obtained in this manner estimate the proportion of times we would falsely reject at least one true null hypothesis if we repeatedly carried out a test according to which we reject all hypotheses with adjusted p-values smaller than some cut-off. Another popular method by Benjamini and Hochberg instead produces adjusted p-values that estimate the false discovery rate, i.e., the expected proportion of true null hypotheses among all hypotheses that are rejected (Benjamini and Hochberg, 1995).

11.3 Advantages and Disadvantages

11.3.1 Advantages

- The methodology described in this chapter starts with a clear definition of what is meant by the impact of a treatment variable A_j on the outcome process $Y(t)$. With this definition in hand, the corresponding parameter can then be estimated separately and directly for each candidate treatment variable. In contrast to other approaches, we do not have to derive these variable importance estimates from a regression of $Y(t)$ on the complete set of treatment variables $A_1, \ldots, A_p$ and potential confounders that was fitted with the goal of accurately predicting $Y(t)$ rather than with the goal of estimating the importance of a particular A_j.
- The definition of variable importance described here is quite flexible since the user can decide which variables are to be included in the adjustment covariates W_j. On one extreme, we can estimate unadjusted variable importance by leaving W_j empty. This would allow us to identify variables that are useful for predicting $Y(t)$ *in our study population*, but not necessarily in a new population of subjects. On the other extreme, we may be able to adjust for all relevant confounders of the relationship between A_j and $Y(t)$, in which case we will obtain estimates of the *causal* effect of A_j on $Y(t)$. In addition, the treatment variables whose impact on $Y(t)$ we would like to estimate can be any extractions of the available baseline covariates. In particular, they can consist of continuous rather than binary variables as in the mutation example considered here. A slightly more complicated example of an interesting extraction is given by cross-product terms like $A_1 \times A_2$ or linear combinations of baseline covariates. Furthermore, we may be interested in the importance of a multivariate treatment variable $\mathbf{A}_j = (A_{j1}, A_{j2})$, which would allow us, for example, to study the impact of the simultaneous presence of two mutations on virologic response.
- The targeted estimation of the impact of A_j on $Y(t)$ allows us in a straightforward manner to obtain measures of statistical significance such as confidence intervals or p-values. Other methods generally do not provide these and thus do not allow us to distinguish between variables whose importance is genuinely different from zero and those whose importance is in fact zero but is estimated to be non-zero due to sampling variation.

- The methodology described here aims to be as robust as possible, i.e., it aims to rely on as few assumptions as possible. In particular, it avoids the assumption of knowing the functional form of the nuisance parameters g and Q *a priori*. Methods that rely on this assumption can yield severely biased estimates of variable importance if the functional form is guessed incorrectly. This is particularly important in the context of modern genomics and proteomics applications in which the number of variables that might be included in these models is very large, making it virtually impossible to guess the correct model.

- We have described three different transformations of the observed data that each give a different estimate of variable importance. This is valuable since these three transformations differ in how they rely on the two nuisance parameters g and Q so that they can be expected to succeed in different situations. In the setting of a clinical trial, for example, the treatment mechanism g is generally known or straightforward to estimate so that variable importance estimates based on the IPTW transformation can be expected to be very reliable.

- Under certain assumptions the variable importance estimates based on the double robust transformation are as precise as possible, meaning, for example, that it would be impossible to obtain more narrow confidence intervals. Specifically, this is the case if both nuisance parameters are estimated reliably and at a fast enough rate (van der Laan, 2006b).

11.3.2 Disadvantages

- Data-adaptive model selection techniques such as the D/S/A algorithm that search through a large space of candidate functional forms are computationally intensive and do not scale well with a growing number of candidate variables to choose from. Hence, it may be necessary to first reduce the number of these candidate variables by, for example, eliminating those that are not associated with $Y(t)$ in univariate regression models. Such variables are unlikely to confound the relationship between the treatment variables and $Y(t)$ and also would add little to the precision with which we could estimate variable importance measures. Alternatively, we may resort to a less exhaustive model selection algorithm such as the one introduced by Kooperberg et al. (1997), at least for the estimation of g where repeated-measures regression is not needed.

- The computational burden of the D/S/A algorithms often makes it difficult to carry out a completely honest bootstrap simulation to estimate the variance of our point estimates. This would require that we repeat the data-adaptive model selection procedure for each bootstrap sample. Due to time constraints, however, we are often forced to treat the selected functional forms for g and Q as given for the purposes of the bootstrap by simply refitting the selected models for each bootstrap sample. This approach ignores the extra variability of our estimates that is introduced by the model selection procedure and thus tends to underestimate their variance somewhat. The p-values obtained in this way are still useful, however, for the purpose of ranking the treatment variables in order of statistical significance. In practice, the variance estimates obtained by ignoring the data-adaptive model selection process are also often not too different from those obtained from a completely honest bootstrap.

- The methodology described here is relatively new and thus has not yet been implemented in the form of a publicly available software package. As described in more detail below, however, the individual steps that are required are fairly straightforward to carry out in a modern statistical computing environment like R (R Development Core Team, 2005).

11.4 Caveats and Pitfalls

The IPTW transformation of the observed data relies crucially on an estimate of the probability that a given subject would have received his or her observed treatment. It weights each observation by the inverse of this probability, thus downweighting observations that were likely to have received their observed treatment and upweighting those that were instead unlikely to have been observed with the treatment we recorded for them. This essentially creates a new sample in which treatment assignment is independent of the baseline covariates, making it straightforward to estimate the impact of treatment A_j on the outcome, *controlling for* W_j, by simply comparing the two groups with $A_j = 0$ and $A_j = 1$. This approach breaks down if for certain values of W_j we never observe one of the two treatment values $A_j = 0$ or $A_j = 1$. In that case we cannot use weighting to create a new sample in which A_j is independent of W_j since the new sample will still not contain any observations with that value of W_j and the missing value of A_j. Variable importance estimates based on the IPTW transformation thus also rely on the so-called *experimental treatment assignment* (ETA) assumption which states that there are no values of W_j for which treatment is assigned in a deterministic fashion. In fact, IPTW-based estimates also perform poorly if the ETA assumption is practically violated, i.e., if for some values of W_j, treatment is assigned in a *nearly* deterministic fashion (Neugebauer and van der Laan, 2005).

We can examine the extent to which the ETA assumption is violated in a number of *ad hoc* ways. We may, for example, look at the proportion of observations for which the probability of having received their observed treatment is very close to zero or one, say less than 0.05 or greater than 0.95. Such observations would hint at values of W_j for which there is very little experimentation with respect to treatment assignment. To look more closely at the relationship between W_j and these fitted probabilities we may also plot the probabilities against the linear combination of W_j that was chosen for the treatment model, or examine observed counts of assigned treatments A_j within deciles of that linear combination.

Current research in this area is investigating variable importance measures that are based on slightly different ideal experiments than those described above (van der Laan, 2006a). Instead of considering hypothetical scenarios in which each member of the population is assigned a particular treatment like $A_j = 1$, these efforts focus on so-called dynamic treatment rules that assign $A_j = 1$ to those members for which this assignment is sensible but $A_j = 0$ to the remaining ones. If for certain treatment histories, for example, it is impossible or very unlikely to observe a particular mutation in the virus of a patient, such rules would never assign such a patient to have this mutation. Treatment rules that are realistic in this sense then no longer rely on the ETA assumption.

The double robust and regression-based transformations rely on an estimate of the regression $E[Y(t) \mid A_j, W_j]$. As mentioned above, we would like to avoid the assumption that the functional form for the dependence of $Y(t)$ on A_j and W_j is known a priori by using data-adaptive model selection techniques. The models selected in this way, however, often contain neither the treatment variable A_j nor any interaction terms between A_j and time, especially in genomics or proteomics applications with a large number of candidate explanatory variables to choose from. Such models are unsatisfactory since they do not allow us to examine the impact of A_j on $Y(t)$ and the change of this impact over time. If we use the regression-based transformation, such models will in fact directly translate into an estimate of zero variable importance for A_j.

To explicitly acknowledge that we are interested in estimating the effect of A_j on $Y(t)$ over time, we might hence fit two separate data-adaptive regression models, one among subjects with $A_j = 0$ and one among subjects with $A_j = 1$. This is problematic, however, for the following reason. Suppose the adjustment variables W_j contain an important confounder that is very strongly correlated with A_j and that has an independent effect on $Y(t)$. Then clearly this variable should be included in a model predicting $Y(t)$ from A_j and W_j to adequately control for confounding by W_j. Within groups defined by A_j, this variable will show very little variation, however, and thus will contribute little to the accurate prediction of $Y(t)$. Model selection procedures are thus unlikely to include this variable in the chosen regression model.

We therefore recommend the following two-step approach: First, fit a data-adaptive regression model for the expectation of $Y(t)$ given W_j alone, excluding A_j from the set of candidate explanatory variables. Then fit a second data-adaptive regression model that is forced to contain all the terms of the first model along with the terms A_j and $A_j \times t$. The first step guarantees that no important confounders are omitted due to strong correlations with A_j. The second step then allows the model selection algorithm to add interaction terms between A_j, $A_j \times t$, and the baseline covariates selected for the first model.

In our description of the data structure we have assumed that all subjects are followed up for the entire duration of the study. This assumption is often not met, with subjects dropping out of the study for various reasons such as moving away or being switched to a new drug regimen due to poor response to the current regimen. The methodology we have described so far will still give reliable estimates of variable importance if such loss to follow-up is not related to what a subject's future outcomes would have been in the two hypothetical scenarios in which each subject is either given the treatment or not. This is, for example, reasonable in the case of subjects moving away since the decision to move away is probably not influenced by what the future outcomes $Y(t)$ might have been. If patients are switched to different drugs due to poor response, however, we will systematically be missing patients that would have had a poor outcome, had it been observed. In the presence of such informative censoring, the estimation procedure described above provides estimates of the importance of A_j among the non-representative subgroup for whom the outcome was measured, which generally differs from its importance in the entire study population.

This problem can be addressed by weighting each observation by the inverse of the probability that the subject was not censored by the time the measurement was made. Such inverse-probability-of-censoring weights work analogously to those used as part of the IPTW transformation in that they artificially create a sample in which censoring is independent of any confounders we would like to adjust for. In practice the needed probabilities can be estimated by modeling the probability of being censored at each time point t given A_j and W_j, using, for example, a Cox proportional hazards model if we treat time as a continuous variable or a pooled logistic regression model if outcome measurements were made at pre-determined intervals for each subject.

An analogous approach can also be taken if the treatment variable A_j or some of the outcome measurements $Y(t)$ are missing for some of the subjects. This could, for example, be the case if some patients never had their genotype measured. As before, we need to address this type of missingness if it is related to what a subject's future outcomes would have been in the two hypothetical scenarios in which each subject is either given the treatment or not. In this case we would use a logistic regression model to estimate the probability that the variable of interest is recorded for a particular subject given what we have observed so far on this person. The observations with available measurements are then weighted by the inverse of these estimated probabilities.

11.5 Alternatives

Many applications in statistics and biology have been concerned with estimating the impact of a number of treatment variables A_j on an outcome process $Y(t)$. Several methods are commonly used for this purpose. Researchers who wish to estimate unadjusted variable importance frequently use generalized estimating equations to regress $Y(t)$ on A_j according to simple models such as

$$E[Y(t) \mid A_j] = \beta_0 + \beta_1 A_j + \beta_2 t + \beta_3 A_j \times t \tag{11.9}$$

The null hypothesis of A_j having no impact on $Y(t)$ is then equivalent to the hypothesis that both β_1 and β_3 are equal to zero, which is straightforward to evaluate using the standard error estimates provided by generalized estimating equations. Conclusions drawn in this way, however, rely on the assumption that the expectation of $Y(t)$ given A_j can in fact be written according to such a simple functional form. If this is not the case, variable importance estimates may be severely biased. This problem becomes even more pressing once we wish to adjust for a number of baseline covariates W_j. If the number of such baseline covariates is large and we include each of them as a simple main-effects term in the model, we will be virtually guaranteed to have mis-specified the functional form according to which the expectation of $Y(t)$ depends on A_j and W_j.

The bias incurred due to such model mis-specification has motivated the use of data-adaptive model selection techniques like the D/S/A algorithm or classification and regression trees (Breiman et al., 1984). In general, researchers include all treatment variables of interest along with the set of potential confounders in the pool of candidate explanatory variables from which the model selection algorithm is then allowed to select a subset of variables for inclusion in the final regression model. In spite of considering a large number of quite complex candidate models, such data-adaptive algorithms frequently end up selecting a model that contains only a relatively small number of covariates. Such models are disappointing for the purpose of estimating variable importance since they do not give us an explicit estimate of the importance of those covariates that are not selected by the algorithm. We can only conclude that these covariates have no impact on $Y(t)$ at all.

This issue is commonly addressed through a resampling-based technique known as bootstrap-**aggregating** or bagging (Breiman, 1996) that is based on re-fitting the data-adaptive regression models on a large number of bootstrap samples drawn from the original data set and then averaging out the coefficient estimates for each variable across all these regression fits. Since different bootstrap samples typically result in the selection of a different set of variables, this approach allows us to obtain non-zero variable importance estimates for a much larger number of variables.

Neither variable importance estimates based on a single data-adaptive regression fit nor those based on bagging lend themselves to an assessment of their statistical significance. This major drawback can be ascribed to the fact that these methods are not designed for the specific purpose of estimating the impact of a number of treatment variables A_j on $Y(t)$ but rather for the purpose of accurately predicting $Y(t)$. Measures of variable importance are only obtained in a secondary step, as a derivative of the estimated regression fit. This is in stark contrast to the methods described in this chapter that are targeted directly at estimating the importance of each separate treatment variable A_j. It is precisely this targeted nature of the variable importance estimates described here that allows us to assess their statistical significance in such a straightforward way. This observation underscores the need to separate the statistical problem of accurately predicting the outcome $Y(t)$ from that of assessing the importance of each of the treatment variables A_j. In many instances we will be interested in the first problem, in which case data-adaptive regression fits obtained, for example, by classification and regression trees represent some of the most powerful tools currently available. If we are interested in estimating variable importance, however, the targeted methods described in this chapter offer many advantages that make them the approach of choice.

11.6 Case Study: HIV Drug Resistance Mutations

In this section we apply the methodology described above to the task of identifying HIV mutations that modulate how well the virus can replicate in the presence of a particular combination of antiretroviral drugs, and thus how well a patient responds to that drug regimen. A considerable number of such drugs are available for treating patients infected with HIV, with the main mechanistic classes consisting of protease inhibitors (PIs), nucleotide and nucleoside reverse transcriptase inhibitors (NRTIs), and nonnucleoside reverse transcriptase inhibitors (NNRTIs). While a patient is being treated with a particular combination of these drugs, the virus frequently acquires a number of mutations that reduce its susceptibility to that drug regimen, requiring the patient to be switched to a new regimen that the virus remains sensitive to. When faced with this situation, clinicians frequently genotype the virus to ascertain the presence or absence of a large number of mutations that are thought to contribute to the resistance to various drugs (Shafer, 2002). This practice motivates us here to identify in a systematic way mutations that have a strong impact on a patient's virologic response to a new drug treatment and that could thus guide a clinician in designing a salvage therapy regimen on the basis of genotypic test results.

The effect of viral mutations on virologic response to therapy can be seriously confounded by a patient's treatment history. Past treatment regimens exert a strong selection pressure on viral evolution, thus affecting the probability that a given mutation is observed. In addition, treatment history can have an independent impact on virologic response by resulting in archived, or latent, virus carrying unobserved mutations that affect response to subsequent treatment regimens. As a result, an unadjusted association observed between a given mutation and treatment response may in fact be due to the presence of other mutations, both observed and unobserved. Treatment strategies vary across populations and evolve over time, potentially resulting in distinct mutation distributions. Thus, control of confounding due to treatment history is needed to ensure that the estimated importance of a given mutation can be more readily generalized to populations other than the original study population.

In order to estimate the causal effect of a given mutation of interest, we would ideally also adjust for the presence of additional mutations. As with treatment history, this would help to ensure that the association we observe between a given mutation and the outcome is indeed causal, rather than due to the effect of other mutations that occur frequently with the mutation of interest. Estimation of such a causal effect is desirable not only from the point of view of mechanistic understanding, but also because it is not dependent on population characteristics such as past treatment patterns that would limit the extent to which it might translate to other HIV-infected populations.

Mutations conferring resistance to drugs of a class different from that targeted by the mutation of interest, thus affecting a distinct viral enzyme, can indeed be controlled for by simply including them in W_j. However, mutations conferring resistance to the same drug class, thus affecting the same viral enzyme, are often correlated to the extent that it is not possible to distinguish which mutation is causally responsible for a given effect. This is due to the fact that, while correlation between mutations affecting distinct viral enzymes occurs primarily as a result of past treatment patterns, correlation between mutations in the same enzyme often occurs as part of an evolutionary pathway towards resistance to drugs targeting that enzyme. Hence, certain mutations are essentially never observed in the absence of another mutation, making it next to impossible to disentangle the individual impacts of these two mutations on virologic response. The statistical consequence of this correlation or collinearity between individual mutations lies in considerable instability of the variable importance estimates we might obtain if we included the other mutations in the same viral enzyme in the group of adjustment variables W_j. Attempting to do so would also cause a severe violation of the ETA assumption since the presence of one mutation might virtually guarantee the presence of the mutation whose importance we are trying to estimate. These considerations suggest that we should not adjust our variable importance estimates for the other mutations in the same viral enzyme.

The data set we use is derived from the Stanford HIV drug resistance database, a patient sample drawn from 16 Kaiser Permanente Northern California clinics for which longitudinal data on HIV reverse transcriptase and protease sequences, antiretroviral treatment, and viral load were recorded. From this database, we identified episodes during which a patient who has failed a previous drug regimen is followed under a new regimen in which at least one of the drugs has been changed. We require that the patient has a baseline viral load measurement available that was taken no more than 24 weeks before initiation of the new treatment. For such records we obtained all viral load measurements that were taken in the 24 weeks following the treatment change. After about 24 weeks, clinicians may switch patients to yet another drug regimen if they do not appear to be responding well to the current salvage therapy regimen. By restricting ourselves to viral load measurements taken before this time point, we avoid having to adjust for the bias introduced into our variable importance estimates by this informative loss to follow-up.

We would like to identify mutations that modulate virologic response to drug regimens that contain the two NRTI drugs lamivudine and stavudine and thus limit ourselves to patients whose salvage therapy regimen contains these drugs. To isolate mutations specific to these two drugs, we exclude patients who are also taking other NRTI drugs. Since mutations thought to confer resistance to NRTI drugs are unlikely to affect susceptibility to PI or NNRTI drugs, we make no requirements as to which drugs of these two classes might be included in the patient's regimen. However, we do control for these covariates in our analyses, as the presence of an NRTI mutation can be associated with the potency of the non-NRTI drugs in the regimen, which in turn can independently affect virologic response. We exclude patients that have never taken an NRTI drug before since they are virtually guaranteed not to have any of the mutations thought to confer resistance to NRTI drugs. Including this group of patients in our analysis would thus cause a severe violation of the ETA assumption. Based on these inclusion criteria, our data set contains 855 viral load measurements from 288 individual treatment change episodes. These measurements were made on 278 individual patients, with a small number of them contributing more than one treatment change episode.

We are interested in assessing the impact on virologic response to lamivudine and stavudine for any mutation in the HIV reverse transcriptase gene that has previously been linked to resistance to NRTI drugs. Mutations are coded as $A_j = 1$ if any of a number of amino acid substitutions potentially related to drug resistance is detected at the given position of the viral enzyme. The mutation 44AD, for example, is considered to be present if either alanine or aspartic acid are found at position 44 of the reverse transcriptase enzyme. For the sake of statistical precision, we only consider mutations that occur at least 15 times among the treatment change episodes we have identified, giving us a total of 14 mutations, whose impact on virologic response we would like to estimate.

We would like to define the outcome $Y(t)$ as the change in log viral load at time t as compared to the baseline measurement made before the treatment change. In our data set, viral loads below $10^{1.7}$ are not detectable so that viral loads below this threshold are simply recorded as below the limit of detection. Since patients whose viral load becomes undetectable during the course of treatment are considered to respond as well as possible to the new drug regimen, we impute the change in log viral load for such patients by the maximal change in log viral load observed across the entire data set, -4.2. We note that this outcome would not be suitable if our goal was to estimate or predict the true change in log viral load resulting from a mutation. However, here, our goal is to estimate the clinical importance of each mutation considered. The outcome definition used thus incorporates the two types of viral response considered a clinical success: 1) A large decrease in viral load, or 2) a final undetectable viral load.

The following variables are used to capture a patient's treatment history: Duration of antiretroviral therapy; number of past regimens; history of past PI, NRTI, and NNRTI drug use; number of PI, NRTI, and NNRTI drugs failed in the past; and history of mono/dual therapy. We characterize the current drug regimen through the total number of drugs as well as the number of PI and NNRTI drugs included in that regimen. Furthermore, we have available information about the duration between baseline viral load measurement, sequencing of the virus, and initiation of the salvage therapy regimen. While we do not adjust each of our variable importance estimates for the presence or absence of the 13 other mutations thought to confer resistance to NRTI drugs, we do adjust them for the presence or absence of a number of mutations that have been linked to resistance to PI and NNRTI drugs. Lastly, we also include Stanford susceptibility scores to the drugs in these two classes that are calculated on the basis of these mutations. This leaves us with a total of 80 baseline covariates that we consider as potential confounders of the relationship between mutations and change in viral load. To reduce the computational burden on the D/S/A algorithm, we reduce the number of baseline covariates to include in W_j by univariate repeated-measures regression of $Y(t)$ on each candidate confounder, only keeping those covariates with adjusted p-values smaller than 0.05. After this initial dimension reduction, the remaining 16 variables considered include the following: The number of past regimens; the number of PI drugs failed in the past; the total number of drugs as well as the number of NNRTI drugs in the new regimen; susceptibility scores for the two NNRTI drugs delavirdine and efavirenz as well as the two PI drugs amprenavir and lopinavir; and four mutations each related to resistance to PI and NNRTI drugs.

We model the treatment mechanisms using the D/S/A algorithm, allowing the algorithm to search through models of up to ten terms, possibly including products comprised of two candidate confounders. The variables most frequently selected for these treatment models include the number of PI drugs failed in the past, the two mutations 90M and 10FIRV that are related to resistance to PI drugs, as well as susceptibility scores for efavirenz and amprenavir. Judging by the percentage of fitted probabilities smaller than 0.05 or greater than 0.95, the majority of mutations appear to satisfy the ETA assumption, with most of these percentages being no greater than 5%. The three notable exceptions to this trend are given by the mutations 75AIMTS, 74IV, and 44AD for which 73%, 48%, and 38%, respectively, of all fitted treatment probabilities are smaller than 0.05 or greater than 0.95. The IPTW-based variable importance estimates for these three mutations may thus be unreliable.

To estimate the expectation of $Y(t)$ given W_j and A_j, we first let the D/S/A algorithm choose an appropriate functional form for predicting $Y(t)$ from W_j. The selected fit includes time t; the susceptibility score for lopinavir; the number of PI drugs failed in the past; the total number of drugs as well as the number of NNRTI drugs in the current regimen; and the mutations 10FIRV, 84AV, and 90M. As described above, we would now like to fit a second data-adaptive regression model that is forced to contain all of these terms along with A_j and $A_j \times t$. In this case, however, we can in fact omit the term A_j based on the following consideration: For 50% of all subjects, baseline viral load was measured within five days, and for 80% of all subjects, it was measured within four weeks of initiation of the new regimen, suggesting that $Y(0)$, the change in log viral load between treatment change and the baseline measurements, is close to zero for the majority of patients. This in turn implies that all variable importance measures should be close to zero at $t = 0$, which makes the term A_j unnecessary.

We described above how we can obtain estimates of the variable importance of each A_j at a chosen set of time points $t_1, \ldots, t_d$. This will result in the simultaneous test of $p \times d$ hypotheses of the form $\Psi_j(t_k) = 0$, where p is the number of treatment variables we are considering. In a first analysis aimed at identifying important explanatory variables for $Y(t)$ rather than examining how their impact on $Y(t)$ changes over time, however, it is often useful to obtain a single summary measure for the variable importance of each treatment variable. This reduces the number of simultaneous hypothesis tests that have to be performed and thus increases the chance of obtaining statistically significant results. In the present case, we can again make use of the assumption that variable importance measures at time $t = 0$ should be close to zero by regressing the transformed observations $D^j_{i,k}$ on time according to the simple model

$$E[D^j(T)] = \beta_j T \tag{11.10}$$

that does not include an intercept term. This functional form is likely to be too simplistic to fit the actual time course of variable importance very well, but we view it more as a means to obtain an interesting summary measure of this time course rather than as an accurate estimate of the time course itself. In particular, we can expect to find a positive coefficient β_j for those mutations that all in all lead to an increase in viral load and a corresponding negative coefficient for those mutations that all in all lead to a decrease in viral load.

As described above, we obtain unadjusted p-values for the hypotheses $\beta_j = 0$ based on a bootstrap estimate of the variance of the estimated coefficient β_j. These p-values are adjusted using the Benjamini-Hochberg method for controlling the false discovery rate. Table 11.1 summarizes the estimated variable importance of each mutation 24 weeks after treatment change corresponding to the estimate obtained for β_j, along with adjusted p-values for the hypothesis that this variable importance is equal to zero. We present estimates based on each of the three different transformations of the data. The mutations are ranked in order of statistical significance according to the estimates based on the double robust transformation.

All three estimators identify the mutation 184IV as having the most significant impact on virologic response to treatment with lamivudine and stavudine. This mutation has in fact been shown to be responsible for high-level resis-

Table 11.1. Variable importance estimates based on the double robust (DR), inverse-probability-of-treatment-weighted (IPTW), and regression-based transformation. Estimates give the impact of a mutation on the change in log viral load after 24 weeks. Mutations marked with * show a significant violation of the ETA assumption.

Mutation	DR	p-value	IPTW	p-value	Regression	p-value
184IV	0.8307	0.0025	0.7200	0.0364	0.6739	0.0004
75AIMTS*	0.5604	0.2611	0.5772	0.4638	0.6255	0.0917
41L	0.3362	0.4187	0.4821	0.1587	0.3017	0.1111
62V	0.6135	0.4187	0.9594	0.1587	0.5534	0.3367
118I	0.3050	0.4658	0.4605	0.2173	0.2631	0.3367
215FY	0.2421	0.5864	0.3877	0.2426	0.2151	0.3367
67EGN	0.1895	0.6566	0.2617	0.4638	0.2378	0.3367
69DN	0.1811	0.8015	0.3349	0.5810	0.1692	0.6081
74IV*	0.2295	0.8015	0.1466	0.6682	0.3208	0.4507
70RGE	0.0931	0.8783	0.1800	0.6221	0.1910	0.4932
210W	0.0800	0.8783	0.0929	0.6781	0.2114	0.3367
219ENQR	− 0.0732	0.8783	− 0.2537	0.5810	− 0.0066	0.9766
44AD*	0.0417	0.9351	0.4866	0.1587	0.1502	0.6416
215CDEIVS	− 0.0041	0.9920	0.2437	0.5810	0.1950	0.6967

tance to lamivudine based on extensive laboratory and clinical data (Boucher et al., 1993; Tisdale et al., 1993; Schurman et al., 1995). Analyses linking HIV mutations directly to *in vitro* drug susceptibility have furthermore identified this mutation as by far the most important mutation conferring resistance to lamivudine (Rhee et al., 2006).

The second most important mutation identified by the double robust and regression-based estimates is given by 75AIMTS. This mutation is ranked much lower based on the IPTW transformation, which as mentioned above, however, can be expected to give unreliable estimates in this case due to a violation of the ETA assumption. 75AIMTS has been shown to confer moderate resistance to the second drug in the regimen we consider, stavudine (Lacey and Larder, 1994). The analyses by Rhee et al. furthermore suggest that this mutation may also be related to drug resistance to lamivudine. The variable importance estimates for the remaining mutations do not approach statistical significance.

Rhee et al. identify 184IV, 69ins, 65R, and 75T as the most important mutations conferring resistance to lamivudine and 69ins, 151M, 77L, 65R, and 75MT as the most important mutations conferring resistance to stavudine. With the exception of 184IV and 75MT, these mutations are not part of our analysis since they are present in fewer than 15 treatment change episodes. Our results that identify 184IV and 75AIMTS as the only important drug resistance mutations for this combination of drugs are hence in excellent agreement with these analyses based on *in vitro* susceptibility tests.

Once important explanatory variables for $Y(t)$ have been identified, it may be of interest to examine in more detail how their impact on $Y(t)$ changes over time. For this purpose, we can estimate the dependence of variable importance on time using data-adaptive or smoothing methods that are better suited to give accurate estimates of this time course than the simple model given in Equation 11.10. Figure 11.1 shows estimates of the time course for 184IV and 75AIMTS based on the loess smoothing technique (Cleveland, 1979). These plots show that 184IV has a sizeable impact on virologic response within a few weeks of treatment initiation, with the effect stabilizing after about ten weeks. The impact of 75AIMTS on virologic response develops somewhat more slowly over time.

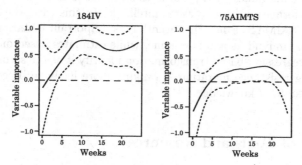

Fig. 11.1. Loess estimates of variable importance over time for the mutations 184IV and 75AIMTS with 95% pointwise confidence intervals.

11.7 Lessons Learned

This data analysis illustrates the importance of choosing an appropriate set of confounders W_j to adjust for when estimating the importance of each treatment variable A_j. In order for estimates to be more likely to translate to populations other than the one the sample was drawn from, one will generally want to adjust for as many of the known confounders of the relationship between A_j and $Y(t)$ as possible. If some of these confounders are collinear with A_j, this will cause the variable importance measures to become very hard to estimate from the data at hand, in which case we may be well-advised not to adjust for these collinear confounders.

The analysis further demonstrates that it is often preferable to obtain low-dimensional summary measures of variable importance time courses for the initial purpose of identifying important explanatory variables for $Y(t)$. At this stage, efforts to estimate variable importance measures at a chosen set of time points, for example, may unnecessarily increase the number of simultaneous hypothesis tests we have to perform and thus lower the chance of obtaining significant results. Once a subset of important explanatory variables has been identified, we may then investigate in more detail how their impact on $Y(t)$ changes over time by using data-adaptive or smoothing methods that make fewer assumptions about the structure of this relationship.

The results obtained here also illustrate the importance of assessing the validity of the ETA assumption if estimates of variable importance are based on the IPTW transformation of the data. Seventy-three percent of all fitted treatment probabilities for the mutation 75AIMTS, for example, are either smaller than 0.05 or greater than 0.95, showing that for a majority of values of W_j it is essentially pre-determined whether a patient will have this mutation or not. In the absence of sufficient variability in the assignment of treatment for all values of W_j, variable importance estimates based on the IPTW transformation become very unreliable. In this case, the IPTW estimates rank 75AIMTS as only the seventh most important mutation conferring resistance to lamivudine and stavudine, while estimates based on the double robust and regression-based transformation identify it as the second most important drug resistance mutation, a ranking more likely to be correct given the current state of knowledge.

11.8 List of Tools and Resources

While the methodology described in this chapter has not yet been implemented in the form of a publicly available software package, the individual steps required as part of the analysis are fairly straightforward to carry out in a modern statistical computing environment like R (R Development Core Team, 2005). Within this environment, repeated-measures regression models based on generalized estimating equations can be fit using the gee() function found in the gee package. A function call of the form

```
gee(Y~X,id=ID,corstr='exchangeable')
```

is used to regress outcome measurements Y, made on individual subjects identified by the ID variable, on a covariate X, assuming an exchangeable correlation structure among the measured outcomes. The D/S/A algorithm is implemented in the DSA package. The function call

```
DSA(X,Y,binind=1,IDlearn=ID,maxsize=10,maxorderint=2,
    maxsumofpow=2)
```

can be used to data-adaptively regress the binary outcome measurements Y, made on individual subjects identified by the ID variable, on a collection of candidate explanatory variables contained in the matrix X. The algorithm will search through models that contain up to 10 terms, including second-order interactions. The sum of powers of the variables contained in any one term cannot exceed two. The polspline package implements the approach by Kooperberg et al. (1997), and offers the polyclass() and polymars() functions for data-adaptive regression of categorical and continuous outcomes, respectively, on a collection of candidate explanatory variables. A simple function call of the form

```
polyclass(Y,X)
```

is used to regress a categorical outcome Y on a collection of candidate explanatory variables contained in the matrix X. Note that these two functions are not suitable for modeling correlated outcomes so that they should only be used for fitting the treatment mechanism. The multtest package offers tools for obtaining valid p-values and confidence intervals in the context of simultaneous hypothesis tests. The Benjamini-Hochberg method for control of the false discovery rate can be carried out by a function call like

```
mt.rawp2adjp(rawp,proc=("BH"))
```

where the vector rawp contains the unadjusted p-values.

The packages gee, polspline, and multtest are available on the R Web site http://www.r-project.org. The DSA package will soon be posted on that Web site as well and can be accessed in the meantime at http://www.stat.berkeley.edu/~laan/Software/.

11.9 Conclusions

Given a list of 14 mutations thought to confer resistance to various NRTI drugs, the data analysis we describe here successfully identifies the mutation 184IV as most useful for predicting virologic response to lamivudine and stavudine. Extensive laboratory and clinical data have previously established 184IV as the most important mutation conferring resistance to lamivudine. The other mutation identified here, 75AIMTS, has been linked to moderate resistance to both lamivudine and stavudine. These results are also in excellent agreement with recent analyses of *in vitro* susceptibility tests and thus illustrate the potential for the variable importance methodology described in this chapter to identify important explanatory variables for a time-varying outcome like viral load.

Acknowledgments

We would like to thank Dr. Robert Shafer and Soo-Yon Rhee from the Stanford HIV drug resistance database for many helpful discussions, as well as kindly making available the data set that was used in our case study analysis.

References

Benjamini, Y. and Hochberg, T. (1995). Controlling the false discovery rate: a practical and powerful approach to multiple testing. *J. Roy. Stat. Soc., Series B*, 85:289–300.

Bland, J.M. and Altman, D.G. (1995). Multiple significance tests: the bonferroni method. *Brit. Med. J.*, 310:170.

Boucher, C.A.B., Cammack, P., Schipper, R., Rouse, P.L., and Cameron, J.M. (1993). High-level resistance to (-) enantiomeric 2'deoxy- 3'thiacytidine (3tc) in vitro due to one amino acid substitution in the catalytic site of human immunodeficiency virus type 1 reverse transcriptase. *Antimicrobial Agents and Chemotherapy*, 37:2231–2234.

Breiman, L. (1996). Bagging predictors. *Machine Learning*, 24(2):123–140.

Breiman, L., Friedman, J.H., Olshen, R.A., and Stone, C.J. (1984). *Classification and regression trees*. The Wadsworth Statistics/Probability series. Wadsworth International Group.

Cleveland, W.S. (1979). Robust locally-weighted regression and smoothing scatterplots. *J. Am. Stat. Assoc.*, 74:829–836.

Dudoit, S. and van der Laan, M. J. (2006). *Multiple Testing Procedures and Applications to Genomics*. Springer. (In preparation).

Kooperberg, C., Bose, S., and Stone, C.J. (1997). Polychotomous regression. *J. Am. Stat. Assoc.*, 92:117–127.

Lacey, S.F. and Larder, B.A. (1994). Novel mutation (v75t) in human immunodeficiency virus type 1 reverse transcriptase confers resistance to 2'-3'didehydro-2',3'-dideoxythymidine in cell culture. *Antimicrobial Agents and Chemotherapy*, 38(6):1428–1432.

Lehmann, E.L. and Romano, J. (2005). *Testing Statistical Hypotheses*. Springer, New York, 3rd edition.

Liang, K. and Zeger, S.L. (1986). Longitudinal data analysis using generalized linear models. *Biometrika*, 73(1):13–22.

McCullagh, P. and Nelder, J. A. (1989). *Generalized linear models (2nd edition)*. London: Chapman & Hall.

Neugebauer, R. and van der Laan, M.J. (2005). Why prefer double robust estimates in causal inference? *J. Stat. Planning and Inference*, 129(1-2):405–426.

R Development Core Team (2005). *R: A language and environment for statistical computing*. R Foundation for Statistical Computing, Vienna, Austria. ISBN 3-900051-07-0.

Rhee, S., Taylor, J., Wadhera, G., Ravela, J., Ben-Hur, A., Brutlag, D., and Shafer, R.W. (2006). Genotypic predictors of human immunodeficiency virus type 1 drug resistance. (Submitted).

Schurman, R., Nijhuis, M., van Leeuwen, R., Schipper, P., de Jong, D., Collis, P., Danner, S.A., Mulder, J., Loveday, C., and Christopherson, C. (1995). Rapid changes in human immunodeficiency virus type 1 rna load and appearance of drug-resistant virus populations in persons treated with lamivu-dine (3tc). *J. Infect. Dis.*, 171:1411–1419.

Shafer, R.W. (2002). Genotypic testing for human immunodeficiency virus type 1 drug restistance. *Clin. Microbiol. Rev.*, 15(2):247–277.

Sinisi, S.E. and van der Laan, M.J. (2004). Deletion/substitution/addition algorithm in learning with applications in genomics. *Stat. Appl. Gen. Mol. Biol.*, 3(1).

Tisdale, M., Kemp, S.D., Parry, N.R., and Larder, B.A. (1993). Rapid in vitro selection of human immunodeficiency virus 1 type 1 resistant to 3'-thyiacytidine inhibitors due to a mutation in the ymdd region of reverse transcriptase. *Proc. Natl. Acad. Sc. USA*, 90:5653–5656.

van der Laan, M.J. (2006a). Causal effects for intention to treat and realistic individualized treatment rules. Technical Report 203, Division of Biostatistics, University of California, Berkeley.

van der Laan, M.J. (2006b). Statistical inference for variable importance. *Intl. J. Biostat.*, 2(1).

Westfall, P.H. and Young, S.S. (1993). *Resampling-based multiple testing: Examples and methods for p-value adjustment*. Wiley, New York.

Zeger, S.L. and Liang, K. (1986). Longitudinal data analysis for discrete and continuous outcomes. *Biometrics*, 42(1):121–130.

Text Mining in Genomics and Proteomics

Robert Hoffmann

Memorial Sloan-Kettering Cancer Center, 1275 York Avenue, New York, NY
10021, USA.
hoffmann@cbio.mskcc.org

12.1 Introduction

With the genome era, biological research has moved from the study of individual genes or proteins to entire biological systems. New high-throughput technologies in biology and medicine (DeRisi et al., 1997; Lander et al., 2001; Phizicky et al., 2003) have led to an explosion in the amount of data and a paradigm shift in biological investigation, such that the bottleneck in research is shifting from data generation to data analysis (Sherlock, 2000). Microarrays and high-throughput protein interaction screens, for instance, often come up with a considerable number of genes whose relevance, both in general and in the studied context, has to be confirmed through manual analysis by experts. These genome-scale experiments confront experimenters with genes or gene products that they might never have heard of.

To make sense of this overwhelming amount of novel data, biologists depend on quick access to previously gathered information on genes, proteins and their interactions in the scientific literature. The sheer number of biomedical publications, however, reaches dimensions that make it difficult to cope with in practice. Every year, half a million new biomedical articles are published, more than thousand a day (NLM, 2006). All in all, results, insights and hypotheses of the past 50 years can be found dispersed over about 15 million scientific papers. Thus, it is impossible for biologists to keep up-to-date – even on a specific subject. In practice this means that the analysis of large-scale data involves jumping back and forth between experimental data and free text searches in PubMed. With these needs arising, methods for mining the biomedical literature are gaining importance in the every day work of biologists.

12.1.1 Text Mining

Text mining refers generally to the computerized process of extracting relevant and non-trivial information and knowledge from free text. The methods

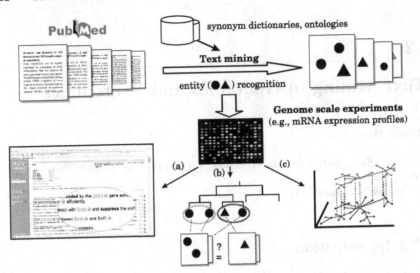

Fig. 12.1. Text mining in genomics and proteomics. Biomedical text is tagged with bio-entities (e.g., genes, MeSH and GO terms) in a high-throughput fashion and applied to the analysis of genome-wide experimental data. (a) Look-up of individual proteins and their relationships in the literature (iHOP (Hoffmann and Valencia, 2005)); (b) Functional coherence analysis of gene groups (Raychaudhuri et al., 2003); (c) Integration of literature derived networks with experimental data (PubGene (Jensen et al., 2006), STRING (von Mering et al., 2005)).

used in text mining are based on information retrieval, data mining, machine learning, statistics and computational linguistics. In biology, information retrieval tools, like the PubMed online repository for biomedical articles, have long been used on a daily basis. The application of automatic information extraction methods to biology is, however, a rather recent development, motivated by the growing interest in systems biology. The experience of the past decade shows that automatic fact extraction is more challenging in molecular biology than in other domains, like economy or newswire. Only very recently, automatic extraction methods are achieving sufficient accuracy to prove useful in biology (Jensen et al., 2006). Important advances have been made in the detection of biomedical entities within scientific text (Fukuda et al., 1998; Proux et al., 1998; Collier et al., 2000; Krauthammer et al., 2000; Friedman et al., 2001; Marcotte et al., 2001; Franzen et al., 2002; Hirschman et al., 2002; Yu et al., 2002; Hanisch et al., 2003; Morgan et al., 2003; Tsuruoka and Tsujii, 2003; Mika and Rost, 2004) and novel ideas for the analysis of large scale experiments have been introduced (Tanabe et al., 1999; Blaschke et al., 2001; Blaschke and Valencia, 2001; Friedman et al., 2001; Masys et al., 2001; Park et al., 2001; Raychaudhuri et al., 2003; Glenisson et al., 2004).

12.1.2 Interactive Literature Exploration

Some of the major difficulties of biomedical literature mining are related to entity recognition and the complexity of sentences in the biomedical domain. For instance, short names and acronyms, which are identical to gene and protein synonyms, but have other meanings (e.g., diseases, methods or therapies) are very common in the biomedical literature. Obviously, if a gene is erroneously identified in the first place, all subsequently extracted facts will be irrelevant or misleading. This is extremely important for the practical relevance of text mining tools, since even small errors can be frustrating, and biologists lose their confidence in automatic extraction methods.

Since many of the important problems have not been solved yet, but demand from the biological domain is steadily increasing, many recent text mining systems step back from the goal of purely automatic information extraction and include the expert user interactively in the analysis (Jenssen et al., 2001; Hoffmann and Valencia, 2004; Glenisson et al., 2004). The *iHOP server* (Information Hyperlinked over Proteins), for example, provides genes and proteins as hyperlinks between sentences and abstracts (Hoffmann and Valencia, 2004) and thus converts the information in PubMed into one navigable network. This way, researchers can move between sentences taken directly from their source abstracts and thus retain control over the reliability of the information they obtain (see Figure 12.1).

In this chapter, I will introduce the basic concepts, caveats and possibilities of current text mining approaches in biology and medicine. I will explore how state of the art text mining methods are used to facilitate the analysis of large scale experimental data. I will discuss complementary methods to analyze expression clusters and the annotation of protein interaction networks. Finally, I will focus on the potential of hyperlinked and navigable text clusters for information retrieval, using the genome-wide study of human chromosome aberrations as a test case.

12.2 Basic Concepts

Many efforts in text mining are related to the fundamental problem of automatically *understanding* natural language and biology serves merely as a demanding challenge for novel methods. Here, I will introduce the basic concepts and terminology of text mining, but with an eye on their relevance to biological problems.

12.2.1 Information Retrieval

Genome-scale experiments, like microarrays, confront experimenters with numerous genes that they might not have heard of previously. One of the first steps in analyzing microarray date involves therefore exhaustive searches in the literature for information on sets of genes. *Information retrieval* (IR) systems are designed to facilitate this process and to identify the textual information (i.e., documents, abstracts or sentences) relevant to a specific topic or query. PubMed, for instance, is a well-known example of an IR system in biology, which provides Boolean queries and document similarity measures to retrieve abstracts upon user request. Boolean queries are the most widespread query type and are also employed by common Internet search engines, like Google, and support the retrieval of documents that include specific combinations of terms and logical operators (e.g., TGFβ AND *pathway*). To efficiently identify and return relevant documents, all IR systems depend on some kind of previously generated indices that define which terms occur in which documents (Witten et al., 1999). In biology, the iHOP system provides one of the most extensive indices of documents for specific proteins and their interactions (Hoffmann and Valencia, 2004, 2005).

12.2.2 Entity Recognition

In molecular biology the detection and identification of entities (e.g., genes and proteins) is the first important step in most text mining approaches. This task, however, has shown to be anything but trivial (Tamames and Valencia, 2006; Hirschman et al., 2002). Biomedical literature is rife with synonyms and short equivocal acronyms for genes, chemical compounds, disorders, methods and therapies. An extreme case is fruit fly research, where collisions with common English words are very frequent (e.g., *white* and *I'm not dead yet* proteins). Moreover, proteins are continuously discovered and described by independent research groups and different symbols and synonyms exist therefore for identical proteins. Finally, the namespace for gene symbols, consisting typically of two to three letters and digits, is limited and collisions between genes with the same synonym are frequent. *Reg1* for example is used as a synonym for four different genes; two in yeast, one in mouse and another in fruit fly. In spite of these difficulties, ongoing research is coming up with increasingly efficient approaches to identify gene or protein synonyms in natural text. Methods are based on dictionaries, rules, and machine learning (Fukuda et al., 1998; Krauthammer et al., 2000; Collier et al., 2000; Jenssen et al., 2001; Morgan et al., 2003; Tsuruoka and Tsujii, 2003; Mika and Rost, 2004; Hoffmann and Valencia, 2005). Systems that use dictionaries for the identification of genes and proteins are most common in practice, because they achieve high precision and make it possible to cross-link identified entities with external databases.

12.2.3 Information Extraction

Information extraction (IE) aims to extract facts of predefined characteristics, like specific relationships between biological entities (e.g., protein interactions, metabolic reactions, phosphorylation events). Naturally, the correct identification of entities (e.g., genes or proteins) is essential for the success of

IE approaches. Problems specific to IE methods include the correct identification of relationships between entities. Currently, there are two complementary approaches available to extract relationships from biomedical texts: *Natural language processing* and *statistical co-occurrence analysis.*

The underlying assumption of co-occurrence analysis is that genes that exhibit a similar pattern of presence and absence over a number of documents or sentences might also function together in a pathway or structural complex. In spite of its simplicity, this method has been successfully applied to a number of biologically relevant problems (Jenssen et al., 2001; Donaldson et al., 2003; Cooper and Kershenbaum, 2005), although subsequent statistical analysis is essential to exclude random or meaningless concurrences.

Co-occurrence, however, is unable to identify the exact entity relationships in complex sentences, which often contain more than two entities and multiple interactions. Moreover, co-occurrence is also not adequate for the extraction of directional relationships. To detect whether the protein *MAPKAPK2* phosphorylates *LSP1* or vice versa, one has to recur to natural language processing methods (NLP). NLP is a subfield of artificial intelligence and linguistics that studies the automated generation and understanding of natural human languages (Hausser, 2001).

12.2.4 Biomedical Text Resources

Over the past decades, Medline has been the most important resource for scientific abstracts from biomedical and other life science journals in electronic format. Medline contains abstracts from more than 4800 journals covering the fields of medicine, nursing, dentistry, veterinary medicine, health, and the preclinical sciences. Currently, the database contains over 12 million citations dating back to 1966 (NLM, 2006).

In 1997, the US National Library of Medicine (NLM) has taken the lead in preserving and maintaining unrestricted access to the electronic literature and started to provide online access to bibliographic information over the PubMed Web portal that includes Medline and OLD-Medline (citations from 1950 through 1965 without abstracts). Moreover, the NLM puts enormous effort into indexing all PubMed citations with MeSH terms (MeSH is the thesaurus of the NLM), publication types, protein accession numbers, and other indexing data. These activities by the NLM have been a major contribution to the development of life sciences and are essential for most text mining efforts in biology.

In recent years most publishing houses started to provide online versions of their journals. Thus, full text in machine-readable form is theoretically becoming available to text mining. However, due to copyright issues, most journals are currently not freely available. This situation is likely to change as the trend towards open access journals is gaining momentum. PubMed Central, for instance, is a novel digital archive that provides standardized and unrestricted access to the available open access literature.

Besides scientific journals, patent databases are a rich source of biomedical information and make their textual information accessible in electronic format. The European Bioinformatics Institute, for example, provides access to biotechnology related abstracts of patent applications (see Resources in Section 12.7).

12.2.5 Assessment and Comparison of Text Mining Methods

The comparison of text mining methods is generally difficult, especially when different text corpora or gold standards are used for training or evaluation. Efforts to evaluate and compare methods systematically are thus crucial for progress of the field. Various efforts have been made by the community to assess and compare different approaches (see Resources Section 12.7). BioCreative, for instance, is an open evaluation of systems on a number of biologically relevant text mining tasks, like the manual curation process behind model organism databases (Hirschman et al., 2005; Blaschke et al., 2005). An important indirect contribution of these assessments is the manual annotation of biological text corpora for training and evaluation. Articles are read by human experts, which highlight relevant biological entities within the text. Such corpora are extremely expensive in their creation and only a few are available (Kim et al., 2003).

12.3   Caveats and Pitfalls

12.3.1 Entity Recognition

Text mining techniques depend on the correct identification of entities such as protein and gene names, chemical compounds, and diseases. This basic step, however, has turned out to be extremely difficult, because of the high semantic overloading of abbreviations and synonyms. The detection of biological entities is hindered by two properties of language well known in the field of text processing: *Synonymy* and *polysemy*. Synonymy refers to the existence of more than one name for the same object, and polysemy indicates that a given term can have multiple meanings. For the retrieval of relevant documents about a specific protein this means that synonymy causes reduced recall (relevant documents are missed). Polysemy again affects precision negatively, due to the retrieval of documents that are not related to the query protein (e.g., documents about a disease or method with the same name). These problems in entity identification have been found to be much harder in biology than the identification of names in areas such as economics or newswire services and standardization of nomenclatures are being demanded from many sides (White et al., 1997). However, despite all efforts to establish nomenclature standards and to assemble dictionaries, official gene names still do not provide a solution to the problem of name detection. In 1994, only 36% of the human genes are mentioned by their official names according to the Human Genome Organization (HUGO) nomenclature, and by 2004 this percentage increases only to about 43% (Tamames and Valencia, 2006). It seems that the dynamics of synonym creation and usage are as vigorous as the evolution of genes and proteins (Hoffmann and Valencia, 2003) and that static nomenclatures and dictionaries will always lag behind. Thus, community efforts to establish a standard vocabulary will probably not succeed unless publishers decide to enforce it, as they have done with the standard deposition of sequences, structures, and expression profiles.

12.3.2 Full Text

A more fundamental limitation is that full text access to the biomedical literature is hampered by copyright restrictions. Most publicly available systems provide therefore mainly information from titles and abstracts. Studies indicate that although abstracts contain the best ratio of keywords per total of words, other sections of the article may be a better source of biologically relevant data (Shah et al., 2003; Liu et al., 2004). When going from abstracts to full text annotation, however, there are a number of issues to be addressed (Schuemie et al., 2004). For instance, full text documents contain cross-references (i.e., anaphoric relationships) between paragraphs and processing full text requires of course more resources. However, addressing full text will be a worthwhile enterprise for next generation systems, even if anaphoric relationships and complex sentences are avoided in an initial phase. Focusing on the extraction of simple relationships (e.g., protein-verb-protein pattern) will maintain a high accuracy, while still increasing the total amount of information in the larger text corpus.

12.3.3 Distribution of Information

With increasing demand for the integration of literature information in the analysis of genome-scale experiments, novel text mining systems have been developed (Jenssen et al., 2001; Raychaudhuri et al., 2002; Hoffmann and Valencia, 2005; Kuffner et al., 2005). However, most of these approaches face the inevitable problem that for a substantial number of genes there is simply no literature information available. This problem is related to the unequal distribution of what is known about individual genes. A few genes are most frequently quoted and attract most attention from the scientific community, while for the rest comparably little has been published. It has been shown that this biased distribution reflects to a large extend the priorities within our society, as is suggested by the most frequently quoted proteins *CD4* and *p53*, which are involved in HIV infection and tumor development (Hoffmann and Valencia, 2003). This bias has important negative effects on the performance of many text mining systems, most importantly on those that rely merely on statistics, e.g., protein co-occurrences in abstracts.

12.3.4 The Impossible

Many problems in biological text mining have been successfully tackled over the past decade, and others, like entity identification, will continuously improve in the near future. There are issues, however, for which no solution is yet in sight. For instance, computational linguists have not yet developed methods that could analyze more than 30% of sentences from PubMed abstracts correctly and transfer the information into structured formal representations (Briscoe and Carroll, 2002). Thus, no computational workflow or data mining strategy should be designed today with dependency on perfect syntactical analysis of extracted sentences. It is however possible to extract information from parts of sentences, which describe for example protein interactions (Ono et al., 2001; Marcotte et al., 2001; Blaschke and Valencia, 2001; Cooper and Kershenbaum, 2005).

12.3.5 Overall Performance

Different methods, training sets and gold standards make it often difficult to choose the appropriate method or Web-based system for a specific problem or the design of bioinformatic workflows. The performance of text mining methods is typically measured by comparison to gold standard data or manual assessment by experts, leading to the estimation of correct retrievals/extractions (true positives, TP), Type I errors (false positives, FP) and Type II errors (false negatives, FN). Precision of a method is then calculated as $TP/(TP + FP)$ and recall or sensitivity as $TP/(TP + FN)$. Within the literature mining community, precision and recall are often combined into a single F-measure:

$$F = \frac{2 \times precision \times recall}{precision + recall} \quad (12.1)$$

According to the recent BioCreative assessment, state of the art systems detect about 80% of protein or gene names in biomedical text (recall) with an accuracy of about 80% (precision) (Hirschman et al., 2005; Blaschke et al., 2005). These results are unsatisfactory, although similar levels of accuracy are in the range of manual curation when assessing the annotation between different curators (Mi et al., 2003). Certainly, limitations of current text mining approaches become apparent in tasks where knowledge extrapolation and interpretation are required (Hirschman et al., 2005). In the next section I will describe how many of these fundamental problems can be circumvented by involving the user interactively in the discovery process (Hoffmann and Valencia, 2004), or by combining text mining information with data from independent sources (e.g., sequence information) (Hoffmann and Valencia, 2004; von Mering et al., 2005).

12.4 Alternatives

Recent advances in high-throughput methods such as microarrays (DeRisi et al., 1997) and protein interaction screens (Phizicky et al., 2003) allow the systematic exploration of functional gene groups, but present a formidable challenge to data analysis. In the absence of structured, computer-readable information about the genes involved, the interpretation of the biological basis of similar gene expression patterns is generally left to the observer. Moreover, the sheer volume of different genes on microarrays and the complexity of all possible ways in which genes might be related physically or functionally complicate the analysis. In practice, the analysis of gene groups derived from genome-scale experiments usually involves switching back and forth between data and literature searches.

At the moment, there are three important complementary approaches in biological text mining that try to make use of literature information to facilitate the interpretation of large scale experimental data (see Figure 12.1). This is, they mimic or support the user in the attempt to read all the documents published about genes in a given cluster and to find properties common to all of them that moreover make sense in the specific context of the experiment.

12.4.1 Functional Coherence Analysis of Gene Groups

First-generation clustering methods have focused on numerical analysis, like unsupervised clustering, and did not incorporate background knowledge about the genes involved. Raychaudhuri et al. (2002, 2003) developed a method called *neighborhood divergence* to quantify the functional coherence of experimentally derived gene groups based on the similarity of documents which mention these genes. This method involves two steps: First, applying hierarchical clustering to a given gene expression data set. Then, resolving hierarchical cluster boundaries to optimize the functional coherence of all clusters. By including literature information in the analysis of gene expression data in this way, Raychaudhuri et al. (2002) make use of functional information when defining expression clusters. For cases where a gene has not been investigated and thus lacks primary literature, articles about well-studied homologous genes can be used.

This strategy has been shown to be successful in identifying biologically coherent gene groups without manual intervention, but has the slight drawback that it does not give direct information on the actual function of a given cluster. A number of similar approaches with varying emphases on the interpretability of term profiles are described in the literature (Shatkay et al., 2000; Masys et al., 2001; Blaschke et al., 2001; Chaussabel and Sher, 2002; Glenisson et al., 2004; Kuffner et al., 2005).

12.4.2 Co-Occurrence Networks

Jenssen et al. (2001) developed a pioneering online system to link co-expression information from a given microarray with a co-citation network constructed from the literature. They employed a straightforward approach to scan ten million biomedical abstracts for gene names and symbols. The existence of co-occurrences of gene symbols in the same abstract was used to build a network of relationships among genes. This approach is based on the assumption that if two entities are repeatedly mentioned together, it is likely that they are somehow functionally related, although the exact kind of relationship remains unknown. Jenssen et al. (2001) chose this simplistic method over detecting particular types of gene-gene relationships to prioritize perspective over detail. This approach is the first to successfully use automated linkages to the literature in assisting the interpretation of array data and represents an important achievement as such. However, the simplicity of the employed text mining method comes at a high price in the analysis of the results. For instance, it is very difficult to verify the importance of specific edges in the co-occurrence network, as the concurring entities might be mentioned in different sentences or in indirect relationships.

This is a general problem of approaches in which natural language is translated into logical or graphical representations. These representations can have manipulative effects on the user's learning process, because they require a specificity that is not achieved by the algorithms employed, e.g., co-occurrences, frames (Blaschke and Valencia, 2001) or regular expressions, etc. Moreover, the graphical representation of large literature networks is in practice rather inappropriate for the analysis and communication of information. The sheer volume of information simply overtaxes most users, and more importantly, the accuracy of the extracted information varies significantly across the network (or any other abstract representation). In practice, this means that the user is forced to check many connections manually, which involves changing back and forth between the text source and the graphical representation. Thus, the creative process of gathering new information and the generation of hypothesis becomes far from intuitive and often frustrating.

12.4.3 Superimposition of Experimental Data to the Literature Network

The iHOP Web site (Hoffmann and Valencia, 2004, 2005) attempts to combine the literature network of concurring genes and proteins with genome-wide experiments in a conceptually similar approach to Jenssen et al. (2001), but without ever leaving the textual representation. In iHOP, every gene has one Web page that contains all the sentences associating it with other genes. Other gene synonyms within sentences serve as hyperlinks to their corresponding Web pages. Thus, each step through the network produces the information pertaining to only a single gene and its associations. In this way, researchers can move between sentences taken directly from their source abstracts and thus retain control over the reliability of the information they obtain. As a byproduct of this novel approach, major difficulties in information retrieval can be mitigated: Ambiguity of synonyms and erroneously identified entities (Hirschman et al., 2002). Most importantly, the underlying gene network remains intact in spite of its representation as hyperlinked text, and it is thus possible to superimpose the iHOP net with network data from other sources. Thereby, a simultaneous exploration of novel and existing knowledge becomes possible. Technically, this is achieved by highlighting sentences in the iHOP network that mention protein associations for which external experimental evidence exists (e.g., from protein interaction screens). Furthermore, this approach is not limited to protein networks, since many biological data sets can be represented as networks. Microarray expression data, for instance, can be transformed into networks, where edges would correspond to gene pairs that exhibit highly correlated expression profiles (Stuart et al., 2003).

12.4.4 Gene Ontologies

A complementary approach that avoids direct queries to the biomedical literature makes use of manually curated and standardized classifications of genes. The Gene Ontology (GO) (Ashburner et al., 2000), for example, allows for simple statistical analysis to check whether co-regulated genes found in a microarray experiment also cluster in the same branches of a classification scheme (gene ontology enrichment analysis (Al-Shahrour et al., 2004; Zeeberg et al., 2005)). This approach, however, depends on complete prior characterization; complete in terms of different genes and the individual functions and roles an individual gene might have during the lifespan of a cell. Although it seems feasible that at some point all genes will be manually classified according to an ontology like GO, it is still a long way until all genes will be described exhaustively. Moreover, one has to keep in mind, that classifications are human attempts to simplify the world, forcing observations into a static perspective. The biomedical literature is much less restricted in this aspect, since it is a collection of mainly independent observations. The disadvantages are obvious: Literature is much more difficult to handle and interpret and often contradicting. However, in contrast to a more rigid classification it contains all described angles on proteins and has thus a higher potential to support novel (previously unclassified and undescribed) functional properties of genes in a given microarray cluster, for instance.

12.5 Case Study

In the following I will discuss three different text mining approaches of increasing complexity. First, I will discuss how to create a simple script-based approach from scratch, which can be used to detect overrepresented terms in

PubMed abstracts. Then I will argue for the use of existing solutions to avoid reinventing the wheel. Finally, I will describe a full-scale text mining solution for extracting comprehensive information on human chromosome aberrations.

Besides the well-known online interface for queries to PubMed, the NCBI also provides a set of programs that provide a stable interface for the Entrez retrieval system. These E-Utilities use a fixed URL syntax that translates a standard set of input parameters into values necessary for various NCBI software components to search and retrieve data from about 20 databases, including abstracts from PubMed. With this, it is possible to develop a simple program that retrieves biomedical abstracts (i.e., in XML) and detects statistically significant terms compared to a background text corpus. The pseudo-code below summarizes the main steps. The same approach can be used to analyze significant genes in any text corpus, although this would involve the lookup of terms in a gene synonym dictionary (e.g., HUGO (White et al., 1997)) or the use of an existing service dedicated to identify biological entities (e.g., genes or diseases) in a given text piece.

```
calculateTermFrequencies(pubmedQuery)
{
  Use E-Utilities to fetch relevant PMIDs for pubmedQuery;
  Use E-Utilities to download abstracts (XML or plain text);
  Split resulting text around blanks;
  Loop through individual terms
      {Homogenize terms, e.g., capitalize;
       (Optional for genes: lookup in gene-synonym dictionary, e.g., from HUGO;
       skip term if not a gene);
       Count occurrence of each term;
      }
  Return term frequencies as hashmap (key=term, value=frequency).
}


main(pubmedQuery)
{
  termFreqBackground = calculateTermFrequencies ("gene", "protein");
  termFreqQuery = calculateTermFrequencies (pubmedQuery);
  Loop through keys (terms) in termFreqQuery as termX
      {Get value for termX (frequency);
       Get frequency for termX from termFreqBackground;
       Compare frequencies for significance (see Mathematical Details);
       Output significant terms.
      }
}
```

Real-world text mining solutions in biology are typically of much higher complexity and expensive on resources. For instance, the detection of biomedical entities (e.g., genes and chemical compounds) within natural text is a computationally very expensive task. The reasons for this are the large number of different genes and gene synonyms, which have to be kept in random access memory to make lookup of terms as time efficient as possible. The iHOP annotation pipeline for example was designed to screen about 12 million abstracts for five million different gene synonyms in a day and runs on a 40 node server cluster (queue system) (Hoffmann and Valencia, 2005). At an average abstract length of 150 words, the total number of examined terms reaches about 1.5 billion, of which each could be one of the five million gene synonyms in the dictionary.

Fortunately, for many basic problems, e.g., tagging or entity recognition, there are ready-made applications available. Many of these systems are available as Web-based systems (see Resources in Section 12.7). Thus, computational biologists should be encouraged to make use of these resources and develop their ideas on top of established services.

The iHOP server, for instance, can be used to retrieve information for specific genes or to retrieve sentences describing a specific protein-protein relationship. Connecting to the iHOP system works similar to the NCBI E-utilities and is URL based, hence the query URL has to encode the protein(s) of interest and the kind of information to be retrieved. Currently, all major database accession numbers are recognized by the iHOP system (e.g., NCBI Gene, UniProt, etc.). Thus, iHOP can be used to provide users of mass spec software or microarray clustering tools with literature information on specific proteins or protein interactions.

In the following I will discuss a full-scale case study aimed at extracting comprehensive information on human chromosome aberrations from biomedical abstracts. In model systems (e.g., mouse or fly) identifying and generating mutations is the usual genetic approach to understanding the role of individual genes. In human populations, natural mutations, such as chromosome aberrations, are a comparable resource for genetic research, since DNA breakage and reciprocal recombination often lead to the fusion or dysregulation of specific genes (Rabbitts, 1994; Heim and Mitelman, 1995; Vogelstein and Kinzler, 2002). Indeed, most human cancers display recurrent chromosome abnormalities. Motivated by this wealth of information, Mitelman et al. (1997) manually collected clinical and morphological data on cancer related chromosome aberrations from the literature. Although the Mitelman database constitutes an important source of detailed clinical information, it depends entirely on expensive manual curation, and contains relatively little molecular information. Here, I describe a text mining approach to generate comprehensive information on all human breakpoints and their relationships to human pathologies. The statistical analysis of this textual information and its combination with genomic data can identify genes directly involved in DNA rearrangements. This case study is well suited to demonstrate the potential of text mining in biology, since it mimics the manual curation process and can thus be directly compared.

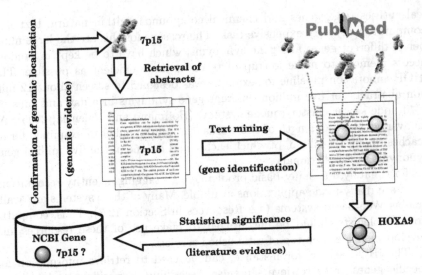

Fig. 12.2. Identification of breakpoint genes in human cells. Abstracts were retrieved from PubMed and clustered by breakpoints (e.g., 7p15). The genes found in a given cluster are not necessarily the actual breakpoint genes; however, the more often a gene is mentioned together with a breakpoint the more likely it is involved in an aberration (literature evidence). False positive associations are eliminated by cross-checking a gene's localization with genomic data (genomic evidence). The final decision on the relevance of a gene is facilitated through a Web-based interface (Hoffmann et al., 2005).

In a first step, PubMed abstracts were screened for references to translocations, insertions and inversions using regular expressions (Friedl, 2002). This identification can be done fully automatically, since aberration codes, for instance t(9;22)(q34;q11.2) are so complex, that they can be easily identified with a negligible error. However, since the online query interface of PubMed does not provide the search for regular expressions, it was necessary to download all relevant abstracts (i.e., abstracts containing the keyword "translocation") and to screen the local copy.

Abstracts were then clustered according to the breakpoint to which they refer. However, each breakage event involves typically two chromosome breakpoints, and it is therefore often unclear to which of the breakpoints a given abstract should be assigned. This is a good example for when entity recognition and information retrieval come to a theoretical limit and only deeper syntactical analysis might be able to uncover the detailed relationships. However, since in many cases not even the author will have the knowledge to select the relevant breakpoint, a computationally expensive syntactical analysis was avoided and weight was put on the subsequent statistical analysis: All abstract clusters were analyzed for the occurrence and frequency of MeSH terms (Kim et al., 2001), associative verbs and genes. The frequencies of these concepts were then compared to their frequencies in a background corpus (consisting of all abstracts referencing chromosome aberrations) to calculate their significance (see Mathematical Details and Figure 12.2).

The premise in the subsequent statistical analysis of this text-mined data is that genes directly affected by recurrent breakage events will be quoted more often in abstracts about the corresponding breakpoint, even if a direct proof for this association has not been described yet. Based on the literature data alone, however, this would result in many false positive gene-breakpoint relationships. It is therefore necessary to integrate the text-mined information with an independently derived source of information. Here, independent evidence is the genomic location of each gene, which can be roughly mapped to specific breakpoints and thus helps to eliminate false positives (see Figure 12.2).

With this approach 343 of 861 literature associated genes were found to localize to recurrent breakpoints. Indeed, for a third of these there is already clear experimental evidence that they are involved in fusion events (Hoffmann et al., 2005).

For the final output through a Web-based application, statistically significant genes and biomedical terms (e.g., disease names) were mapped back onto their source sentences, where they serve as hyperlinks between different parts of each breakpoint cluster. Diseases and associative verbs are also highlighted and hyperlinked within the text to further facilitate the perception of associations with human pathologies (Hoffmann et al., 2005).

12.6 Lessons Learned

Since one aim of this case study is to partially reproduce the manually generated Mitelman database (Mitelman et al., 1997) it is intriguing to directly compare the text mining strategy with the manual approach. As can be expected, the text mining approach outperforms the manually maintained database, both in terms of screened papers and detected genes (Hoffmann et al., 2005). However, the degree of details found in the manually curated database is unrivaled by the automatic approach and not only because of the limited access to full texts. The Mitelman database provides detailed facts on clinical cases, including sex and age of patients, tissue types and morphologies. To extract all these bits of information and most importantly the correct relationships between them is a task for which no out-of-the-box solution exists and which stresses the limits of current NLP methods (Briscoe and Carroll, 2002; Jensen et al., 2006). The lower degree of details can be compensated, however, by the larger amount of processed documents and the subsequent statistical analysis. For example, the smaller amount of data in the manually curated database did not allow for detecting which gene was actually involved in the aberration at a given breakpoint.

Moreover, a large number of genes and diseases that can be found in abstracts about chromosome aberrations were filtered or ignored by the human curators. Thus, whereas manual curation often depends on the opinion of human experts, text mining approaches are quite objective. Most importantly, it is very difficult to recover the exact literature evidence that underlies manual curation. Automatic text mining systems, on the contrary, transform free text information into a structured representation, which can be easily linked to the original source texts and moreover allows for integrating other structured information, like information from external databases, such as UniProt, Gene and OMIM.

Although this case study involved an exhaustive screen of about 12 million PubMed abstracts for thousands of gene synonyms and MeSH terms, it is in principle only a "blow-up" of the pseudo-code example above. For the computationally most expensive step in this study (i.e., entity detection) I used the iHOP annotation pipeline (Hoffmann and Valencia, 2005). However, this step could also be implemented with varying complexity as described in Jenssen et al. (Jenssen et al., 2001) or by making use of one of the other existing entity recognition tools (see Resources in Section 12.7). Hence, text mining methods are becoming ready to deliver and straightforward to use or implement in the design of novel analysis tools and workflows.

12.7 List of Tools and Resources

Text mining in biology is a very young and dynamic field and the following list of services, tools and methods can only provide a static view. However it gives an useful overview about what is possible today and what kind of biological questions are being addressed (see Tables 12.1, 12.2, 12.3).

12.8 Conclusion

With the expected improvement of experimental high-throughput technologies, amount and quality of genome-wide data will increase continuously in the near future. Instead of single microarray experiments, for instance, multiple experiments will become common practice as well as systematic comparisons with standardized data resources. For the final interpretation, however, human expertise will continue being essential to integrate background knowledge and to formalize novel hypotheses. This task is extremely demanding in itself, and even more so since much of the necessary background information is scattered across the literature. This is the operational area of text mining in biology in the 21st century.

Therefore, any text mining method in biology should be assessed according to its contribution to this endeavor. Here I have discussed exciting and complementary text mining approaches that are becoming useful in the analysis

Table 12.1. Text mining tools and resources in biology – *information retrieval.*

Source	Description and URL
PubMed	The National Library of Medicine's search service. http://www.pubmed.org
NCBI E-utilities	Access to Entrez data outside of the regular Web query interface. http://www.ncbi.nlm.nih.gov/entrez/query/static/eutils_help.html
E-BioSci	The European platform for access and retrieval of full text and factual information in the Life Sciences. http://www.e-biosci.org
iHOP	Retrieves relevant sentences for protein interactions and protein function. Provides the network of concurring proteins for navigating the biomedical literature. http://www.ihop-net.org/
EBI Literature DBs	Biotechnology related abstracts of patent applications from the European Patent Office. http://www.ebi.ac.uk/Databases/literature.html
Google Scholar	Google search and ranking of scholarly literature in full text format and citation information. http://scholar.google.com
GoPubMed	Classifies and highlights PubMed query results according to the Gene Ontology. http://www.gopubmed.org
MedMiner	Filters and organizes sentences in the literature based on a gene, gene-gene or gene-drug query. http://discover.nci.nih.gov/textmining
EBIMed	Web application that combines Information Retrieval and Information Extraction from Medline. http://www.ebi.ac.uk/Rebholz-srv/ebimed
XplorMed	Organizes PubMed queries according to the MeSH ontology and summarizes content. http://www.ogic.ca/projects/xplormed
Textpresso	Information retrieval and extraction system for biological literature of C. elegans. http://www.textpresso.org
CrossRef	Full text search. http://www.crossref.org/crossrefsearch.html
PubMed Central	Digital archive of biomedical and life sciences journal literature in full text. http://www.pubmedcentral.org
HighWire Press	Repository of free, full-text, peer-reviewed content. http://highwire.stanford.edu

Table 12.2. Text mining tools and resources in biology – *information extraction.*

Source	Description and URL
BioIE	Rule-based system that extracts informative sentences from PubMed query results. `http://umber.sbs.man.ac.uk/dbbrowser/bioie/`
JournalMine	Queries the biomedical literature for specific entity relationships. `http://textmine.cu-genome.org/gridsphere/gridsphere`
iProLINK	Protein annotation and tagging. `http://pir.georgetown.edu/iprolink`
PubGene	Gene-to-gene co-citation network that can be used for microarray analysis. `http://www.pubgene.org`
KAT	Annotate proteins from scientific references. `http://www.bork.embl-heidelberg.de/kat/index.html`
Data integration	
TxtGate	Summarization and analysis of groups of genes based on text. `http://tomcat.esat.kuleuven.be:8080/txtgate/home.jsp`
STRING	Integration of protein interaction extracted from the literature with information from complementary methods. `http://string.embl.de`
Entity recognition	
ABNER	Entity detection. `http://www.cs.wisc.edu/~bsettles/abner`
GAPSCORE	Protein gene name tagger. `http://bionlp.stanford.edu/gapscore`
NLProt	Protein/gene name tagger. `http://rostlab.org/services/nlprot/`
Protein interactions	
Chilibot	Relationship extraction tool. `http://www.chilibot.net/`
PreBIND	Data mining tool that helps researchers locate biomolecular interaction information in the scientific literature. `http://prebind.bind.ca`
Knowledge discovery	
Arrowsmith	A tool for identifying links between two sets of PubMed articles. `http://arrowsmith.psych.uic.edu`
BITOLA	Aims to facilitate the discovery of potentially new relations between biomedical concepts. `http://www.mf.uni-lj.si/bitola`
HCAD	Provides comprehensive information on human chromosomal aberrations, including genes and disease relationships. `http://www.ihop-net.org/UniPub/HCAD/`
G2D	Finds literature links between OMIM entries and genes from a specific chromosomal location. `http://www.ogic.ca/projects/g2d_2`

Table 12.3. Text mining tools and resources in biology – *annotated text corpora.*

Source	Description and URL
BioCreative corpus	Corpus of protein annotation relevant text.
	`http://www.pdg.cnb.uam.es/BioLINK/`
FetchProt	`http://fetchprot.sics.se`
GENETAG	`ftp://ftp.ncbi.nlm.nih.gov/pub/tanabe`
GENIA	Annotated corpus related to human blood transcription factors.
	`http://www-tsujii.is.s.u-tokyo.ac.jp/GENIA`
PennBioIE	`http://bioie.ldc.upenn.edu`
Yapex	`http://www.sics.se/humle/projects/prothalt`
Assessments	
BioCreative Challenge	Text mining of protein names and annotations.
	`http://www.pdg.cnb.uam.es/BioLINK/`
	`BioCreative.eval.html`
KDD challenge	Information extraction of Drosophila gene expression information.
	`http://www.biostat.wisc.edu/~craven/kddcup/`
	`tasks.html`
TREC Genomics track	IR, document classification and question answering.
	`http://ir.ohsu.edu/genomics/`
Part-of-speech taggers	Marking up the words in a text with their corresponding parts of speech (e.g., verbs, nouns).
Brill	`http://www.cs.jhu.edu/~brill`
TNT	`http://www.coli.uni-saarland.de/~thorsten/tnt`
TreeTagger	`http://www.ims.uni-stuttgart.de/~schmid`
Nat. Language Parsers	Derive the grammatical structure of sentences, e.g., which groups of words are units (phrases) and which words are the subject or object of a verb.
CASS	`http://www.vinartus.net/spa`
Collins Parser	`http://people.csail.mit.edu/mcollins`
Stanford Parser	`http://nlp.stanford.edu/software`

of genome-wide data: Methods to asses the coherence of gene groups (Raychaudhuri et al., 2003), to integrate experimental data with literature networks (Jenssen et al., 2001; Hoffmann and Valencia, 2005; von Mering et al., 2005) and ways to make a simultaneous analysis of literature and experimental data possible (Hoffmann and Valencia, 2004). Many of these methods have led to the development of tools and Web sites ready to use. Other promising methods are still in an experimental phase, but will soon reach production-state. Hence, developers of novel analysis software and workflows are able to choose from a variety of stable text mining solutions. The recent development in science towards open and freely accessible full text-resources will further catalyze this progress.

However, I have also discussed some of the important difficulties and caveats that text mining methods are still facing in biology: The vast number of ambiguous acronyms and symbols and the complexity of scientific language. Addressing these problems at a pragmatic level is important, but it cannot be an aim on its own. Not in the light of gigabytes of data to be expected from large-scale experiments in the near future. Text mining in biology has to focus on biology-driven problems to maintain the momentum gained over the past decade. Thus, some problems that are due to the complexity of natural language might be neglected, but these deficits will be more than compensated by the integration with independently derived sources of information (e.g., large scale experimental data and *in silico* predictions), as pioneered in recent years by a number of groups (Jenssen et al., 2001; Hoffmann and Valencia, 2004; von Mering et al., 2005). Following this direction, text mining in biology will live up to its full potential and will become an integral element of all future approaches to analyze and interpret novel data.

12.9 Mathematical Details

To assess the content of a given document cluster, one can compare the frequencies of scientific terms within the cluster to their frequencies in a reference cluster (e.g., all documents). The probability (P_T) of finding a term (T) the observed number of times (k) in a document cluster (C) is then calculated from the Poisson distribution, given the known reference frequency (p) and the total number of terms in the cluster (n). This approximation is valid when the total number of terms in the reference cluster is much greater than n and p is small.

$$P_T(k\,|n,p) = e^{-np}\frac{(np)^k}{k!} \tag{12.2}$$

where $n \in N$, the number of terms assigned to a document cluster (C),
$k = 1, 2, ...n$, the number of occurrences of term (T) within the cluster (C),
p is the relative frequency of term (T) in the reference cluster. In practice the log probability can be calculated to avoid floating point errors and $n!$ can be estimated using Stirling's approximation for large n:

$$\ln P_T(k|n,p) = -np + k\ln(np) + k - k\ln(k) \tag{12.3}$$

References

Al-Shahrour, F., Diaz-Uriarte, R., and Dopazo, J. (2004). FatiGO: A web tool for finding significant associations of gene ontology terms with groups of genes. *Bioinformatics*, 20(4):578–580.

Ashburner, M., Ball, C.A., Blake, J.A., Botstein, D., Butler, H., Cherry, J.M., Davis, A.P., Dolinski, K., Dwight, S.S., Eppig, J.T., Harris, M.A., Hill, D.P., Issel-Tarver, L., Kasarskis, A., Lewis, S., Matese, J.C., Richardson, J.E., Ringwald, M., Rubin, G.M., and Sherlock, G. (2000). Gene ontology: Tool for the unification of biology. the gene ontology consortium. *Nat. Genet.*, 25(1):25–29.

Blaschke, C., Leon, E. A., Krallinger, M., and Valencia, A. (2005). Evaluation of BioCreAtIvE assessment of task 2. *BMC Bioinformatics*, 6 Suppl. 1.

Blaschke, C., Oliveros, J. C., and Valencia, A. (2001). Mining functional information associated with expression arrays. *Functional and Integrative Genomics*, 1(4):256.

Blaschke, C. and Valencia, A. (2001). The potential use of SUISEKI as a protein interaction discovery tool. *Genome informatics series: Proc. Workshop on Genome Informatics*, 12:123.

Briscoe, T. and Carroll, J. (2002). Robust accurate statistical annotation of general text. *Proc. 3rd Intl. Conf. Language Resources and Evaluation*, pages 1499–1504.

Chaussabel, D. and Sher, A. (2002). Mining microarray expression data by literature profiling. *Genome Biol*, 3(10):RESEARCH0055.

Collier, N., Nobata, C., and Tsujii, J. (2000). Extracting the names of genes and gene products with a hidden markov model. *Proc. COLING 2000*, pages 201–207.

Cooper, J.W. and Kershenbaum, A. (2005). Discovery of protein-protein interactions using a combination of linguistic, statistical and graphical information. *BMC Bioinformatics*, 6(1):143.

DeRisi, J.L., Iyer, V.R., and Brown, P.O. (1997). Exploring the metabolic and genetic control of gene expression on a genomic scale. *Science*, 278(5338):680–686.

Donaldson, I., Martin, J., de Bruijn, B., Wolting, C., Lay, V., Tuekam, B., Zhang, S., Baskin, B., Bader, G.D., Michalickova, K., Pawson, T., and Hogue, C.W. (2003). Prebind and textomy–mining the biomedical literature for protein-protein interactions using a support vector machine. *BMC Bioinformatics*, 4:11.

Franzen, K., Eriksson, G., Olsson, F., Asker, L., Liden, P., and Coster, J. (2002). Protein names and how to find them. *Int. J. Med. Inf.*, 67(1–3):49–61.

Friedl, J.E.F. (2002). *Mastering regular expressions*. O'Reilly, Sebastopol, 2nd edition.

Friedman, C., Kra, P., Yu, H., Krauthammer, M., and Rzhetsky, A. (2001). GENIES: A natural-language processing system for the extraction of molecular pathways from journal articles. *Bioinformatics*, 17 Suppl. 1:S74–82.

Fukuda, K., Tamura, A., Tsunoda, T., and Takagi, T. (1998). Toward information extraction: Identifying protein names from biological papers. *Pac. Symp. Biocomput.*, pages 707–718.

Glenisson, P., Coessens, B., Van Vooren, S., Mathys, J., Moreau, Y., and De Moor, B. (2004). Txtgate: profiling gene groups with text-based information. *Genome Biol.*, 5(6):R43.

Hanisch, D., Fluck, J., Mevissen, H. T., and Zimmer, R. (2003). Playing biology's name game: Identifying protein names in scientific text. *Pac. Symp. Biocomp.*, pages 403–14.

Hausser, R.R. (2001). *Foundations of Computational Linguistics: Human-Computer Communication in Natural Language.* Springer, Berlin/New York, 2nd edition.

Heim, S. and Mitelman, F. (1995). *Cancer Cytogenetics.* Wiley-Liss, New York, 2nd edition.

Hirschman, L., Morgan, A.A., and Yeh, A.S. (2002). Rutabaga by any other name: Extracting biological names. *J Biomed Inform*, 35(4):247–59.

Hirschman, L., Yeh, A., Blaschke, C., and Valencia, A. (2005). Overview of biocreative: Critical assessment of information extraction for biology. *BMC Bioinformatics*, 6 Suppl. 1.

Hoffmann, R., Dopazo, J., Cigudosa, J. C., and Valencia, A. (2005). HCAD, closing the gap between breakpoints and genes. *Nucleic Acids Res.*, 33(Database issue):D511–D513.

Hoffmann, R. and Valencia, A. (2003). Life cycles of successful genes. *Trends Genet.*, 19(2):79–81.

Hoffmann, R. and Valencia, A. (2004). A gene network for navigating the literature. *Nat. Genet.*, 36(7):664.

Hoffmann, R. and Valencia, A. (2005). Implementing the iHOP concept for navigation of biomedical literature. *Bioinformatics*, 21 Suppl. 2:ii252–ii258.

Jensen, L.J., Saric, J., and Bork, P. (2006). Literature mining for the biologist: from information retrieval to biological discovery. *Nat. Rev. Genet.*, 7(2):119–129.

Jenssen, T.K., Laegreid, A., Komorowski, J., and Hovig, E. (2001). A literature network of human genes for high-throughput analysis of gene expression. *Nat. Genet.*, 28(1):21–28.

Kim, J.D., Ohta, T., Tateisi, Y., and Tsujii, J. (2003). GENIA corpus–A semantically annotated corpus for bio-textmining. *Bioinformatics*, 19 Suppl. 1:I180–I182.

Kim, W., Aronson, A.R., and Wilbur, W.J. (2001). Automatic MeSH term assignment and quality assessment. *Proc. AMIA Symp.*, pages 319–23.

Krauthammer, M., Rzhetsky, A., Morozov, P., and Friedman, C. (2000). Using BLAST for identifying gene and protein names in journal articles. *Gene*, 259(1–2):245–252.

Kuffner, R., Fundel, K., and Zimmer, R. (2005). Expert knowledge without the expert: Integrated analysis of gene expression and literature to derive active functional contexts. *Bioinformatics*, 21 Suppl. 2:ii259–ii267.

Lander, E.S., Linton, L.M., and Birren, B., et al. (2001). Initial sequencing and analysis of the human genome. *Nature*, 409(6822):860–921.

Liu, F., Jenssen, T.K., Nygaard, V., Sack, J., and Hovig, E. (2004). FigSearch: A figure legend indexing and classification system. *Bioinformatics*, 20(16):2880–2882.

Marcotte, E.M., Xenarios, I., and Eisenberg, D. (2001). Mining literature for protein-protein interactions. *Bioinformatics*, 17(4):359–363.

Masys, D.R., Welsh, J.B., Lynn Fink, J., Gribskov, M., Klacansky, I., and Corbeil, J. (2001). Use of keyword hierarchies to interpret gene expression patterns. *Bioinformatics*, 17(4):319–326.

Mi, H., Vandergriff, J., Campbell, M., Narechania, A., Majoros, W., Lewis, S., Thomas, P. D., and Ashburner, M. (2003). Assessment of genome-wide protein function classification for *Drosophila melanogaster*. *Genome Res.*, 13(9):2118–2128.

Mika, S. and Rost, B. (2004). Protein names precisely peeled off free text. *Bioinformatics*, 20 Suppl. 1:I241–I247.

Mitelman, F., Mertens, F., and Johansson, B. (1997). A breakpoint map of recurrent chromosomal rearrangements in human neoplasia. *Nat. Genet.*, 15 Spec. No.:417–474.

Morgan, A., Hirschman, L., Yeh, A., and Colosimo, M. (2003). Gene name extraction using FlyBase resources. *ACL-03 Workshop on Natural Language Processing in Biomedicine*, pages 1–8.

NLM (2006). Yearly citation count totals. *US National Library of Medicine*. http://www.nlm.nih.gov.

Ono, T., Hishigaki, H., Tanigami, A., and Takagi, T. (2001). Automated extraction of information on protein-protein interactions from the biological literature. *Bioinformatics*, 17(2):155–161.

Park, J.C., Kim, H.S., and Kim, J.J. (2001). Bidirectional incremental parsing for automatic pathway identification with combinatory categorial grammar. *Pac. Symp. Biocomp.*, pages 396–407.

Phizicky, E., Bastiaens, P.I., Zhu, H., Snyder, M., and Fields, S. (2003). Protein analysis on a proteomic scale. *Nature*, 422(6928):208–215.

Proux, D., Rechenmann, F., Julliard, L., Pillet, V.V., and Jacq, B. (1998). Detecting gene symbols and names in biological texts: A first step toward pertinent information extraction. *Genome Inform. Ser. Workshop Genome Inform.*, 9:72–80.

Rabbitts, T.H. (1994). Chromosomal translocations in human cancer. *Nature*, 372(6502):143–149.

Raychaudhuri, S., Chang, J.T., Imam, F., and Altman, R.B. (2003). The computational analysis of scientific literature to define and recognize gene expression clusters. *Nucleic Acids Res.*, 31(15):4553–4560.

Raychaudhuri, S., Schutze, H., and Altman, R.B. (2002). Using text analysis to identify functionally coherent gene groups. *Genome Res.*, 12(10):1582–1590.

Schuemie, M.J., Weeber, M., Schijvenaars, B.J., van Mulligen, E.M., van der Eijk, C.C., Jelier, R., Mons, B., and Kors, J.A. (2004). Distribution of information in biomedical abstracts and full-text publications. *Bioinformatics*, 20(16):2597–2604.

Shah, P.K., Perez-Iratxeta, C., Bork, P., and Andrade, M.A. (2003). Information extraction from full text scientific articles: Where are the keywords? *BMC Bioinformatics*, 4:20.

Shatkay, H., Edwards, S., Wilbur, W. J., and Boguski, M. (2000). Genes, themes and microarrays: using information retrieval for large-scale gene analysis. *Proc. Intl. Conf. Intell. Syst. Mol. Biol.*, 8:317–328.

Sherlock, G. (2000). Analysis of large-scale gene expression data. *Curr. Opin. Immunol.*, 12(2):201–205.

Stuart, J.M., Segal, E., Koller, D., and Kim, S.K. (2003). A gene-coexpression network for global discovery of conserved genetic modules. *Science*, 302(5643):249–255.

Tamames, J. and Valencia, A (2006). The success (or not) of HUGO nomenclature. *Genome Biology*, in press.

Tanabe, L., Scherf, U., Smith, L.H., Lee, J.K., Hunter, L., and Weinstein, J.N. (1999). MedMiner: An internet text-mining tool for biomedical information, with application to gene expression profiling. *Biotechniques*, 27(6):1210–1214, 1216–1217.

Tsuruoka, Y. and Tsujii, J. (2003). Boosting precision and recall of dictionary-based protein name recognition. *ACL-03 Workshop on Natural Language Processing in Biomedicine*, pages 1–8.

Vogelstein, B. and Kinzler, K.W. (2002). *The Genetic Basis of Human Cancer*. McGraw-Hill Medical Pub. Division, New York, 2nd edition.

von Mering, C., Jensen, L.J., Snel, B., Hooper, S.D., Krupp, M., Foglierini, M., Jouffre, N., Huynen, M.A., and Bork, P. (2005). STRING: Known and predicted protein-protein associations, integrated and transferred across organisms. *Nucleic Acids Res.*, 33(Database issue):D433–437.

White, J.A., McAlpine, P.J., Antonarakis, S., Cann, H., Eppig, J.T., Frazer, K., Frezal, J., Lancet, D., Nahmias, J., Pearson, P., Peters, J., Scott, A., Scott, H., Spurr, N., Talbot, C., Jr., and Povey, S. (1997). Guidelines for human gene nomenclature (1997). HUGO Nomenclature Committee. *Genomics*, 45(2):468–471.

Witten, I.H., Moffat, Alistair, and Bell, Timothy C. (1999). *Managing gigabytes: Compressing and indexing documents and images*. Morgan Kaufmann Series in Multimedia Information and Systems. Morgan Kaufmann Publishers, San Francisco, Calif., 2nd edition.

Yu, H., Hatzivassiloglou, V., Rzhetsky, A., and Wilbur, W. J. (2002). Automatically identifying gene/protein terms in medline abstracts. *J. Biomed. Inform.*, 35(5–6):322–330.

Zeeberg, B.R., Qin, H., and Narasimhan, S. et al. (2005). High-throughput GoMiner, an 'industrial-strength' integrative gene ontology tool for interpretation of multiple-microarray experiments, with application to studies of Common Variable Immune Deficiency (CVID). *BMC Bioinformatics*, 6:168.

Index